SpringerBriefs in Electrical and Computer Engineering

Signal Processing

Series editors

Woon-Seng Gan, Singapore
C.-C. Jay Kuo, Los Angeles, USA
Thomas Fang Zheng, Beijing, China
Mauro Barni, Siena, Italy

More information about the series at http://www.springer.com/series/11560

Gang Yu · Junsong Yuan
Zicheng Liu

Human Action Analysis with Randomized Trees

 Springer

Gang Yu
Junsong Yuan
School of Electrical and Electronic
  Engineering
Nanyang Technological University
Singapore
Singapore

Zicheng Liu
Microsoft Research
Redmond, WA
USA

ISSN 2196-4076
ISBN 978-981-287-166-4
DOI 10.1007/978-981-287-167-1

ISSN 2196-4084 (electronic)
ISBN 978-981-287-167-1 (eBook)

Library of Congress Control Number: 2014946770

Springer Singapore Heidelberg New York Dordrecht London

Printed on acid-free paper

Springer is part of Springer Science+Business Media (www.springer.com)

Preface

Human action analysis plays a critical role in human-centered computing, as in many applications we require to analyze and understand human actions in the big video data, such as video surveillance, health care, and human–computer interaction. In this book, we focus on the human action analysis techniques based on randomized trees, which can handle the complexity and variations of human actions, and also provide efficient action analysis in large-scale video data. It covers both supervised and unsupervised randomized trees. When there are sufficient amount of labeled data available, supervised randomized trees provide a fast method for spacetime interest point matching. When the amount of labeled data is limited such as in example-based action search, unsupervised randomized trees can be used to leverage the unlabeled data. The goal of this book is to provide a comprehensive overview of the supervised and unsupervised randomized trees and their applications to four different tasks in human action analysis: action classification, action detection, action search, and action prediction. For students who are interested in human action analysis or randomized trees, this book provides a good guide to quickly bring the students to the research front. For researchers who have been working in this area, we hope this book provides a useful reference.

We start with a brief review of previous work on the four tasks of applications of human action analysis. Chapter 2 describes supervised randomized trees which improve both the matching accuracy and the spacetime search efficiency. Chapter 3 presents unsupervised randomized trees which are suitable for exemplar-based action search where the amount of labeled data is extremely small. Chapter 4 introduces a propagative Hough voting technique that leverages spatiotemporal contextual information in addition to the unlabeled data. Action prediction is described in Chap. 5. Finally we conclude in Chap. 6.

Acknowledgment

We would like to thank Norberto A. Goussies for his contribution to this book. This book is supported in part by Nanyang Assistant Professorship SUG M4080134 and Microsoft Research Gift Funding.

Contents

Chapter 1
Introduction to Human Action Analysis

Abstract Human action analysis plays a critical role in human centered computing, as in many applications we require to analyze and understand human actions in the big video data, such as video surveillance, health care and human–computer interaction. In this chapter, we start from an overview of human action analysis problems, and then discuss the recent development of the four human action analysis applications as well as tree-based approaches.

Keywords Action recognition · Action detection · Action search · Action prediction · Tree-based approaches

1.1 Overview

With the recent development of video technologies, cameras, especially surveillance cameras, are widely used in many places and environments such as shops, airports, streets, and so on. However, the functions of these cameras are not fully utilized due to the expensive manual labor and slow manual speed. In this book, we decide to explore the great potentials of the video cameras: analyzing the human actions and activities.

Human action analysis is of great importance to our daily lives. For instance, in a supermarket, it is better to send off an alarm while someone is stealing rather than after the stealing, because it can possibly prevent this criminal activity and also provide more time for the security guard to react. In this book, we focus on four important human action analysis tasks: human action recognition, human action detection, human action search, and human action prediction.

In the past decade, there have been a lot of great works in the human action recognition [16, 22, 30]. The goal of human action recognition is to determine the action category of a segmented video clip, referred to as X, among a set of classes $\{1, \cdots, K\}$. However, usually, the testing videos are not segmented, i.e., we need to perform spatio-temporal localization at the same time of action classification. Thus, for the action detection [7, 22, 40, 47], it aims to not only determine the category but also localize the position of the human action, i.e., $f^c(X) \rightarrow [x, w, y, h, t, d]$, where

© The Author(s) 2015
G. Yu et al., *Human Action Analysis with Randomized Trees*,
SpringerBriefs in Signal Processing, DOI 10.1007/978-981-287-167-1_1

$f^c(X)$ refers to the detection function for the action category c and $[x, w, y, h, t, d]$ refers to the spatio-temporal position of the human action (center position x, y, t and scale w, h, d).

Action search [45, 46] is the task to retrieve, from a database of videos, the actions which have the same meaning as the given query videos. Similar to action detection, the output is a 3D volume with spatio-temporal center (x, y, t) and scale (w, h, d). The difficulty for human action search is that the number of query videos is usually very small, and as a result, there are not enough labeled videos to learn a discriminative model. In addition, we cannot make any assumptions on the classes of actions that people will retrieve.

Action prediction [27, 44] is even more challenging because we need to efficiently predict the action given incomplete observations. Compared with action detection, action prediction focuses more on determining whether or not a specified action is happening. Thus, it does not require the exact localization of the human actions. In this book, instead of determining the exact 3D-volume of the human actions, we focus on determining the existence of an action given incomplete observations for the action prediction problem. Formally, we want to find a function:

$$f(O) \rightarrow \{0, 1, \cdots, K\}, \tag{1.1}$$

where $O \subset X$ refers to the incomplete observations, and $\{1, \cdots, K\}$ refers to the category of the predicted action while 0 means no target action is happening. Observation O can be either on segmented videos similar to action classification or unsegmented videos similar to action detection. In this book, we only focus on predicting actions with segmented videos, but our algorithm can be extended to handle the cases where the testing videos are not segmented.

Figure 1.1 summarizes the differences among human action recognition, action detection, action search, and action prediction (action and activity are used interchangeably). In the following section, we will give more detailed discussions on human action recognition, action detection, action search, and action prediction.

	Action Recognition	Action Detection	Action Search	Action Prediction
Testing Video X				
Description	Segmented	Un-segmented	Un-segmented	Segmented or Un-segmented Incomplete observations
Training f(X)	A set of positive and negative videos	A set of positive and negative videos	One positive query video	A set of positive and negative videos
Objective	$f(X) \rightarrow \{1, \dots K\}$	$f^c(X) \rightarrow [x, w, y, h, t, d]$	$f^c(X) \rightarrow [x, w, y, h, t, d]$	$f(O) \rightarrow \{0, 1, \dots K\}$ Where $O \subset X$

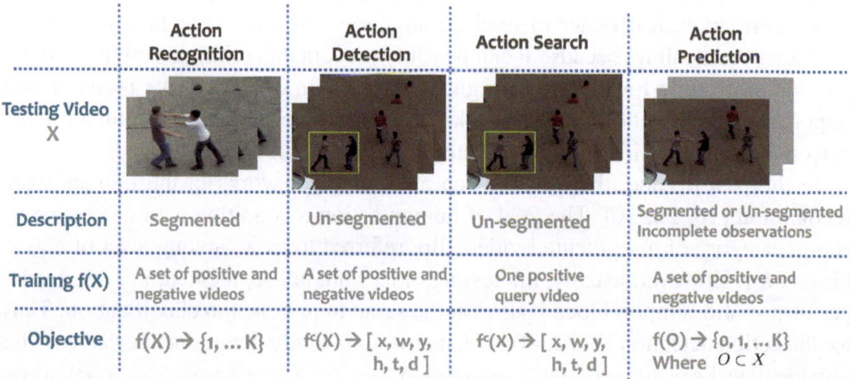

Fig. 1.1 Comparison of action recognition, action detection, action search, and action prediction

1.2 Action Recognition

Based on the successful development of video features, e.g., STIP [27], cuboids [15], and 3D HoG [26], many human action recognition methods have been developed. Previously, [1, 9, 18, 41] rely on the human detection or even human pose estimation for action analysis. But human detection, tracking, and pose estimation in uncontrolled environments are challenging problems.

Without relying on auxiliary algorithms such as human detection and human tracking, [5, 19, 25, 33] perform action recognition by formulating the problem as a template matching process. In [25], an extended canonical correlation analysis-based video-to-video similarity measure is proposed with good performance on both action recognition and action detection. In [5], it learns a spatial-temporal graph model for each activity and classifies the testing video as the one with the smallest matching cost. In [19], temporal information is utilized to build the "string of feature graph." Videos are segmented at a fixed interval and each segment is modeled by a feature graph. By combining the matching cost from each segment, they can determine the category of the testing video as the one with the smallest matching cost. In [33], similar idea of partitioning videos to small segments is used but video partition problem is solved with dynamic programming.

There are several limitations in these matching-based algorithms. First, the template matching algorithms, e.g., graph matching [5, 19], are computationally intensive. For example, in order to use these template-based methods for activity localization, we need to use sliding-windows to scan all the possible subvolumes, which is an extremely large search space. Despite the fact that [33] can achieve fast speed, the proposed dynamic BoW-based matching is not discriminative since it drops all the spatial information from the interest points. Similarly, in [5, 19], the temporal models are not flexible to handle speed variations of the activity pattern. Third, a large training dataset will be needed to learn the activity model in [5, 33]. However, in certain action recognition applications, the amount of training data is extremely limited.

1.3 Action Detection

Usually, the testing videos are not spatially-temporally segmented, i.e., background and noisy information exist in the videos. One example is shown in Fig. 1.1. In order to locate the desired actions, the direct way is to slide all the candidate windows in the search space and find those with largest matching score [13, 25, 34, 50]. The intensive computational cost for sliding window makes it impractical for many real world applications.

In [49], 3D branch and bound search is proposed to speed up the localization with mutual information-based similarity score. Cao et al. [7] further improves 3D branch and bound search [49] based on the Maximum a Posterior (MAP) estimation framework for adaptive action detection. In [8], multiple spatial-temporal interest point features have been combined based on a Gaussian Mixture model representation. Based on the similar GMM model, [37] proposes a Hierarchical Filtered Motion (HFM) method to recognize and detect actions in crowded videos.

The performance of 3D branch and bound search [49] is superior but the computational cost is still quite intensive for high-resolution videos and large datasets. In [20], it presents speed-up techniques like Top-K and spatial-downsampling based on 3D branch and bound search, which can significant reduce the computational cost for multiple instance action detection. Hough voting is an alternative way for action localization. In [18], it utilized Hough forest to vote a few candidates and evaluate the similarity between the training videos and testing candidates. However, Hough voting is heavily affected by the dimensionality of the Hough space, which will fail when we vote for the spatio-temporal subvolume. Thus, human detection and tracking may be needed to perform the spatial localization for Hough voting-based approaches [18].

Other interest work on action detection include [38] and [42]. In [38], max-margin structured SVM is proposed for spatio-temporal action localization. Yao et al. [42] presents animated pose templates for detecting short-term, long-term, and contextual actions from cluttered scenes in videos.

1.4 Action Search

Despite great successes in action recognition and detection, action retrieval, on the other hand, is less exploited. We can roughly categorize most of the existing action retrieval algorithms into two classes based on the number of query samples. Algorithms in the first category [14, 31] perform the sliding window search on the database with a single query sample. The idea for both [14] and [31] is to represent query and database videos with some features and to compare the similarity based on query-to-subvolume measurements. In [14], visual space-time oriented energy measurements are used while a five-layer hierarchical space-time model is employed in [31]. One limitation of these techniques is that with a single query sample, it is challenging to model action variations. Besides, an action retrieval system usually involves user interactions but their approaches do not have the capability to incrementally refine their models based on the user feedback. The other category of action retrieval algorithms, for example [28], is based on a set of query samples, usually including both positive and negative samples. Despite the fact that they work well in uncontrolled videos, the computational cost is high and they would fail if insufficient number of query samples are provided. Apart from the above work, there exist some other algorithms in the literature. For example, [2, 10, 17] rely on auxiliary tools like storyboard sketches, semantic words, and movie transcripts for action retrieval, while [39] is specifically focused on quasi-periodic events. Ikizler-Cinbis et al. [23] performs action retrieval based on static images. Ji et al. [24] retrieves the similar human action patterns with spatiotemporal vocabulary.

1.5 Action Prediction

The problem of human activity prediction has been proposed in [33]: *inference of the ongoing activity given temporally incomplete observations.* Integral bag-of-words (BoW) and dynamic bag-of-words are proposed in [33] to enable activity prediction with only partial observations. Despite certain successes of [33], it still has several limitations. First, since the BoW model ignores the spatial-temporal relationships among interest points, it is not discriminative enough to describe human activities. Also, although integral BoW and dynamic BoW in [33] consider the temporal information by matching between subintervals, there lacks a principled way to determine the optimal interval length. Finally, as we usually have a large number of categories of activities to detect, it demands an algorithm whose computational complexity is sublinear or constant to the number of categories. Besides, in [21], a novel max-margin formulation is proposed which can well handle the early event detection problem. However, the training for different ratio of observations are independent for [21], which significantly increased the training cost. Cao et al. [6] presents a novel approach based on a posterior maximization formulation. Sparse coding is employed to compute the activity likelihood at each segment and a global likelihood is computed by combining each segment.

1.6 Tree-Based Approaches

Tree-based approaches will be the central component for local interest point matching in all the proposed algorithms of this book. Based on the matching results, our approaches can perform various applications like action detection and search. Traditionally, there are two categories of random trees, which are commonly used for computer vision and machine learning communities: supervised random trees and unsupervised random trees.

Supervised random trees are originally proposed in [4] as a classifier. Depends on the applications, different split criteria can be used. In Chap. 2, a minimum classification rate-based random forest has been presented with good performance on action classification. In Chap. 5, multiclass balanced trees are proposed to make a trade-off between the discriminative ability and balance of the trees. In [3], it utilizes random forest and ferns for image classification. Other work that leverage supervised trees include [18, 29, 35]. In [18], random forest is used to cluster the local interest points for voting. Similar idea has been utilized in [35]. In [29], an extensive study of node splitting in random forest construction is discussed. Besides, a comprehensive reference of decision forest for computer vision can be referred to [11].

Unsupervised randomized trees [48], however, have not been explored as much as the supervised ones in computer vision. Since it does not need any supervised information, it can be applied to different applications. For example, it can be used for approximate nearest neighbor search as in [36], manifold learning [12], clustering [32], and semantic hashing [43]. In Chaps. 3 and 4, we utilize different unsupervised random trees for action search and recognition.

1.7 Outline of the Book

This book proposes a few solutions for the human action analysis problems which are of great importance to our daily lives. Specifically, we mainly focus on four challenging tasks: human action recognition, human action detection, human action search, and human action prediction. The related works with these topics have been briefly discussed. The remaining chapters are organized as follows:

- Chapter 2 presents an action recognition and detection algorithm with efficient spatio-temporal localization. Supervised fandom forest has been utilized to train the model and our algorithm is significantly faster than the state-of-art action detection algorithm [49] with little compromise of the accuracy.
- Chapter 3 describes an action search algorithm based on a few query samples. Since the computational cost is an important factor for online interaction system, our algorithm targets at reducing the computational cost based on the sliding-window framework. With the help of unsupervised random trees and coarse-to-fine branch and bound search, our algorithm can finish a search of 5-h database in only a few seconds.
- Chapter 4 introduces a Hough voting-based approach for human action recognition and action search. Different from the algorithms in Chaps. 2 and 3, spatio-temporal configuration of local interest points has been considered to improve the detection. Besides, random projection trees have been used to leverage the underlying data distribution, which can relieve the problem of limited training data and improve the matching accuracy.
- Chapter 5 presents a simple yet surprisingly effective solution for human activity prediction problem. Based on the similar framework as in Chap. 4, we propose Multi-class Balanced Random Forest, which significantly outperforms the state-of-arts in the human action prediction problem.
- Chapter 6 concludes the book and discusses potential directions for the future work.

Table 1.1 summarizes the potential applications for the proposed algorithm in each chapter.

Table 1.1 The targeted applications for each chapter

	Action recognition	Action detection	Action search	Action prediction
Chapter 2	√	√		
Chapter 3	√		√	
Chapter 4	√	√	√	
Chapter 5	√			√

References

1. M. R. Amer, S. Todorovic, A chains model for localizing participants of group activities in videos, in *ICCV* (2011)
2. Y. Aytar, M. Shah, J. Luo, Utilizing semantic word similarity measures for video retrieval, in *CVPR* (2008)
3. A. Bosch, A. Zisserman, and X. Munoz, Image classification using random forests and ferns, in *Proceedings of the IEEE International Conference on Computer Vision*, pp. 1–8 (2007)
4. L. Breiman, Random forests. Mach. Learn. **45**, 5–32 (2001)
5. W. Brendel, S. Todorovic, Learning spatiotemporal graphs of human activities, in *ICCV* (2011)
6. Y. Cao, D. Barrett, A. Barbu, S. Narayanaswamy, H. Yu, A. Michaux, Y. Lin, S. Dickinson, J. Siskind, S. Wang, Recognizing human activities from partially observed videos, in *CVPR* (2013)
7. L. Cao, Z. Liu, T.S. Huang, Cross-dataset action recognition, in *Proceedings of the IEEE Conference on Computer Vision and Pattern Recognition*, pp. 1998–2005 (2010)
8. L. Cao, Y.L. Tian, Z. Liu, B. Yao, Z. Zhang, T.S. Huang, Action detection using multiple spatio-temporal interest point features, in *IEEE Conference on Multimedia Expo*, pp. 340–345, (2010)
9. W. Choi, S. Savarese, Learning context for collective activity recognition, in *CVPR* (2011)
10. J. Collomosse, G. McNeill, Y. Qian, Storyboard sketches for content based video retrieval, in *ICCV* (2009)
11. A. Criminisi, J. Shotton, *Decision Forests for Computer Vision and Medical Image Analysis* (Springer, Berlin, 2013). ISBN 978-1-4471-4929-3,
12. S. Dasgupta, Y. Freund, Random projection trees and low dimensional manifolds. ACM symposium on theory of computing (STOC), pp. 537–546 (2008)
13. K.G. Derpanis, M. Sizintsev, K. Cannons, R.P. Wildes, Efficient action spotting based on a spacetime oriented structure representation, in *CVPR* (2010)
14. K.G. Derpanis, M. Sizintsev, K. Cannons, R.P. Wildes, Efficient action spotting based on a spacetime oriented structure representation, in *IEEE Conference on Computer Vision and Pattern Recognition*, pp. 1990–1997 (2010)
15. P. Dollar, V. Rabaud, G. Cottrell, S. Belongie, Behavior recognition via sparse spatio-temporal features. Workshop on Visual Surveillance and Performance Evaluation of Tracking and Surveillance (2005)
16. A. Efros, A. Berg, G. Mori, J. Malik, Recognizing action at a distance, in *ICCV*, pp 726–733 (2003)
17. A. Gaidon, M. Marszalek, C. Schmid, Mining visual actions from movies, in *BMVC* (2009)
18. J. Gall, A. Yao, N. Razavi, L. Van Gool, V. Lempitsky, Hough forests for object detection, tracking, and action recognition, in *IEEE Transaction on Pattern Analysis and Machine Intelligence*, pp. 2188–2202 (2011)
19. U. Gaur, Y. Zhu, B. Song, A. Roy-Chowdhury, a string of feature graphs model for recognition of complex activities in natural videos, in *ICCV* (2011)
20. N.A. Goussies, Z. Liu, J. Yuan, Efficient search for top-K video subvolumes for multi-instance action detection, in *IEEE International Conference on Multimedia and Expo* (2010)
21. M. Hoai, F. DelaTorre, Max-margin early event detectors, in *CVPR* (2012)
22. Y. Hu, L. Cao, F. Lv, S. Yan, Y. Gong, T.S. Huang, Action detection in complex scenes with spatial and temporal ambiguities, in *Proceedings of IEEE International Conferebce on Computer Vision*, pp. 128–135 (2009)
23. N. Ikizler-Cinbis, R.G. Cinbis, S. Sclaroff, Learning actions from the web, in *ICCV* (2009)
24. R. Ji, H. Yao, X. Sun, S. Liu, Actor-independent action search using spatiotemporal vocabulary with appearance hashing. Pattern Recogn. **44**, 624–638 (2011)
25. T.-K. Kim, R. Cipolla, Canonical correlation analysis of video volume tensors for action categorization and detection, in *PAMI* (2009)
26. A. Klaser, M. Marszalek, A spatio-temporal descriptor based on 3D-gradients, in *BMVC* (2008)

27. I. Laptev, On space-time interest points. Int. J. Comput. Vis. **64**(2–3), 107–123 (2005)
28. I. Laptev, P. Prez, Retrieving actions in movies, in *Proceedings of the ICCV* (2007)
29. X. Liu, M. Song, D. Tao, Z. Liu, L. Zhang, J. Bu, C. Chen, Semi-supervised node splitting for random forest construction, in *CVPR* (2013)
30. R. Messing, C. Pal, and H. Kautz, Activity recognition using the velocity histories of tracked keypoints, in *IEEE International Conference on Computer Vision*, pp. 104–111 (2009)
31. H. Ning, T.X. Han, D.B. Walther, M. Liu, T.S. Huang, Hierarchical space-time model enabling efficient search for human actions. IEEE TCSVT **19**, 808–820 (2009)
32. D. Nister, H. Stewenius, Scalable recognition with a vocabulary tree, in *CVPR* (2006)
33. M.S. Ryoo, Human activity prediction: early recognition of ongoing activities from streaming videos, in *ICCV* (2011)
34. H.J. Seo, P. Milanfar, Action recognition from one example. IEEE Trans. Pattern Anal. Mach. Intell. **33**, 867–882 (2010)
35. J. Shotton, A. Fitzgibbon, M. Cook, T. Sharp, M. Finocchio, R. Moore, A. Kipman, A. Blake, Real-time human pose recognition in parts from single depth images, in *CVPR* (2011)
36. C. Silpa-Anan, R. Hartley, Optimised KD-trees for fast image descriptor matching, in *CVPR* (2008)
37. Y. Tian, L. Cao, Z. Liu, Z. Zhang, Hierarchical filtered motion for action recognition in crowded videos. IEEE Trans. Syst. Man Cybern. C, **42**(3), 313–323 (2012)
38. D. Tran, J. Yuan, Max-margin structured output regression for spatio-temporal action localization, in *Neural Information Processing Systems* (2012)
39. P. Wang, G.D. Abowd, J.M. Rehg, Quasi-periodic event analysis for social game retrieval, in *International Conference on Computer Vision* (2009)
40. Y. Xie, H. Chang, Z. Li, L. Liang, X. Chen, D. Zhao, A unified framework for locating and recognizing human actions, in *CVPR* (2011)
41. B. Yao, L. Fei-Fei, Modeling mutual context of object and human pose in human-object interaction activities, in *CVPR* (2010)
42. B. Yao, B. Nie, Z. Liu, S.C. Zhu, Animated pose templates for modeling and detecting human actions, in *IEEE Transactions on Pattern Analysis and Machine Intelligence* (2013)
43. G. Yu, J. Yuan, Scalable forest hashing for fast similarity search, in *IEEE International Conference on Multimedia and Expo* (2014)
44. G. Yu, J. Yuan, Z. Liu, Predicting human activities using spatio-temporal structure of interest points, in *ACM Multimedia* (2012)
45. G. Yu, J. Yuan, Z. Liu, Real-time human action search using random forest based hough voting, in *ACM Multimedia* (2011)
46. G. Yu, J. Yuan, Z. Liu, Unsupervised random forest indexing for fast action search, in *CVPR* (2011)
47. G. Yu, A. Norberto, J. Yuan, Z. Liu, Fast action detection via discriminative random forest voting and top-K subvolume search. IEEE Trans. Multimedia **13**(3), 507–517 (2011)
48. G. Yu, J. Yuan, Z. Liu, Action search by example using randomized visual vocabularies. IEEE Trans. Image Process. **22**(1), 377–390 (2013)
49. J. Yuan, Z. Liu, Y. Wu, Discriminative video pattern search for efficient action detection. IEEE Trans. Pattern Anal. Mach. Intell. **33**, 1728–1743 (2011)
50. J. Yuan, Z. Liu, TechWare: video-based human action detection resources. Sig. Process. Mag. IEEE **27**(5), 136–139 (2010)

Chapter 2
Supervised Trees for Human Action Recognition and Detection

Abstract Efficient action detection in unconstrained videos is as challenging problem due to the cluttered backgrounds and the large intra-class variations of human actions. In this chapter, we characterize a video as a collection of spatio-temporal interest points, and locate actions via searching for spatio-temporal video subvolumes of the highest mutual information score toward each action class. A random forest is constructed to efficiently generate discriminative votes from individual interest points, and a fast top-K subvolume search algorithm is developed to find all action instances in a single round of search. Without significantly degrading the performance, such a top-K search can be performed on down-sampled score volumes for more efficient localization. Experiments on a challenging MSR Action Dataset II validate the effectiveness of our proposed multiclass action detection method.

Keywords Action detection · Top-K subvolume search · 3D Branch-and-bound search · Spatial downsampling · Random forest voting

2.1 Introduction

According to the literature, human action recognition and detection have been widely exploited. Template matching-based approach [4, 12] and tracking-based approach [14] work well in some constrained environment. However, there are still two major challenges to address.

First of all, for the template matching method, only a single template is usually provided to perform action recognition and detection [4, 12]. In such a case, a single template cannot well characterize the intra-class variations of an action and is not discriminative enough for classification. Second, different from object detection, the search space in the spatio-temporal video space is extremely large. It thus greatly increases the computational cost for these template-based approaches. For example, it is very time-consuming to search actions of different spatial scales and different temporal durations in the video space. Although the recently proposed spatio-temporal branch-and-bound search method [16, 18] can significantly improve the search speed, it is still not fast enough to handle high-resolution videos (e.g., 640 ×

© The Author(s) 2015   9
G. Yu et al., *Human Action Analysis with Randomized Trees*,
SpringerBriefs in Signal Processing, DOI 10.1007/978-981-287-167-1_2

480 and higher). Considering that the spatio-temporal localization is computationally more demanding for high resolution videos, it is important to provide efficient solutions for high-resolution videos. Moreover, given a video dataset containing multiple action instances, it is desirable to efficiently detect all of them in one round of search.

To address the above challenges in action recognition and detection, we propose a random forest-based template matching method, as shown in Fig. 2.1. Without performing background subtraction or human body tracking, each video sequence is characterized by a collection of spatio-temporal interest points (STIPs). During the training phase, random forest is constructed to leverage the distribution of the STIPs from both positive and negative classes in the high-dimensional feature space. During the testing phase, each individual point matches the query class through the pre-built random forest, and provides an individual voting score toward each action type. Following the mutual information maximization formulation in [16], action detection becomes finding the spatio-temporal video subvolume with the maximum mutual information score.

Compared with the nearest neighbor-based matching scheme in [16], our proposed random forest-based approach enables a much more efficient interest point matching without degrading the matching quality. Meanwhile, as both positive and negative action samples are taken into account while building the random forest, our proposed method not only handles intra-class action variations well, but also provides more discriminative matching to detect action instances. To reduce the computational overhead in searching high resolution videos, we improve the original spatio-temporal branch-and-bound search method in [16] on two aspects. First, instead of performing branch-and-bound search in the original score volume, we propose to search a down-sampled score volume for efficient action localization. Our theoretical analysis shows that the error between the optimal solution of the down-sampled volume and that of the original volume can be upper bounded. Second, we propose a top-K search method to enable the detection of multiple action instances simultaneously in a single round of branch-and-bound search. It provides an efficient solution for multiclass multiple instance action detection.

To evaluate the efficiency and generalization ability of our proposed method, we perform a cross-dataset action detection test: our algorithm is trained on the KTH dataset and tested on the MSR action dataset II, which contains 54 challenging video sequences of both indoor and outdoor scenes. The extensive multiclass action detection results show that, ignoring the feature extraction cost, our proposed method can search a one-hour 320×240 video sequence in less than half an hour. It can detect actions of varying spatial scales, and can well handle the intra-class action variations including performing style and speed variations, and even partial occlusions. It also can handle cluttered and dynamic backgrounds. The proposed Top-K volume search algorithm is general and can be used for any other applications of video pattern search.

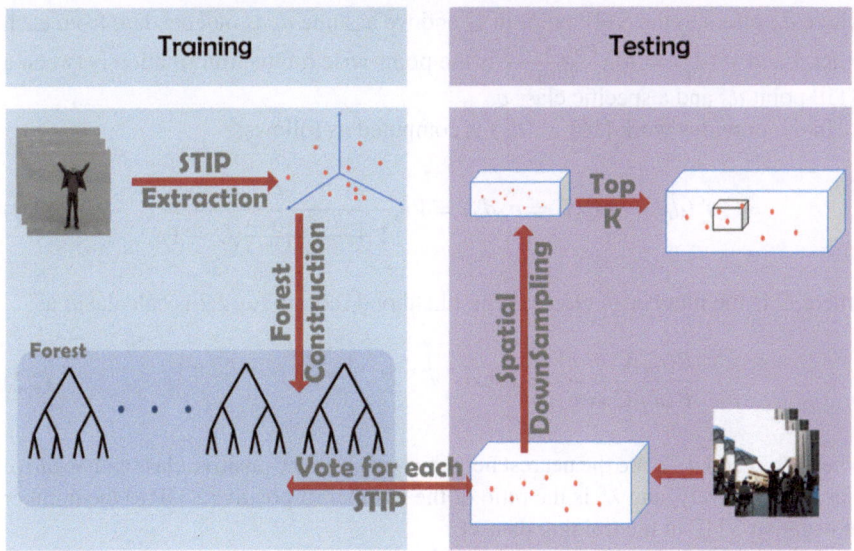

Fig. 2.1 The overview of our random forest-based video subvolume search

2.2 Multiclass Action Recognition

2.2.1 Mutual Information-Based Classification

We represent an action as a collection of spatio-temporal interest points (STIP) [7], where $d \in \mathbb{R}^N$ denotes an N-dimensional feature vector describing a STIP. The reasons to use STIP feature is its superior performance for human action recognition and detection. Besides, the descriptor for STIP, i.e., HOG and HOF, is suitable for our random forest-based framework. A comparison of different detectors and descriptors can be seen in [13]. Denote the class label set as $\mathcal{C} = \{1, 2, \ldots, C\}$.

In order to recognize different action classes, we evaluate the pointwise mutual information between a testing video clip $\mathcal{Q} = \{d_q\}$ and one action class $c \in \mathcal{C}$:

$$
\begin{aligned}
\mathrm{MI}(\mathbf{C} = c, \mathcal{Q}) &= \log \frac{P(\mathcal{Q}|\mathbf{C}=c)}{P(\mathcal{Q})} \\
&= \log \frac{\prod\limits_{d_q \in \mathcal{Q}} P(d_q|\mathbf{C}=c)}{\prod\limits_{d_q \in \mathcal{Q}} P(d_q)} \\
&= \sum_{d_q \in \mathcal{Q}} \log \frac{P(d_q|\mathbf{C}=c)}{P(d_q)},
\end{aligned}
\tag{2.1}
$$

where d_q refers to the STIP point in $\mathcal{Q}$ and we assume d_q is independent from each other. Each $s^c(d_q) = \log \frac{P(d_q|\mathbf{C}=c)}{P(d_q)}$ is the point-wise mutual information between a STIP point d_q and a specific class c.

In the previous work [16], $s^c(d_q)$ is computed as follows:

$$s^c(d_q) = \mathrm{MI}(\mathbf{C} = c, d_q) = \log \frac{C}{1 + \frac{P(d_q|\mathbf{C} \neq c)}{P(d_q|\mathbf{C}=c)}(C-1)}, \tag{2.2}$$

where C is the number of classes. The likelihood ratio in Eq. 2.2 is calculated as:

$$\frac{P(d_q|\mathbf{C} \neq c)}{P(d_q|\mathbf{C} = c)} \approx \lambda^c \exp^{-\frac{1}{\sigma^2}(\|d_q - d_{NN}^{c-}\|^2 - \|d_q - d_{NN}^{c+}\|^2)}, \tag{2.3}$$

where d_{NN}^{c+} and d_{NN}^{c-} are the nearest neighbors of d_q in the positive class and negative class, respectively, and λ^c is the ratio of the number of positive STIP to the number of negative STIP in the training dataset.

Despite its good performance, Eq. 2.3 has two limitations:

- In order to calculate the likelihood ratio in Eq. 2.3, we need to search the nearest neighbors d_{NN}^{c+} and d_{NN}^{c-}. Although locality sensitive hash (LSH) has been employed for fast nearest neighbor search, it is still time-consuming for large high dimensional dataset.
- Only two STIPs are used to approximate the likelihood ratio in Eq. 2.3, which is not accurate.

To address the two problems, we reformulate the voting score $s^c(d_q)$ in Eq. 2.2 as:

$$\begin{aligned} s^c(d_q) = \mathrm{MI}(\mathbf{C} = c, d_q) &= \log \frac{P(d_q|\mathbf{C}=c)}{P(d_q)} \\ &= \log \frac{P(\mathbf{C}=c, d_q)}{P(\mathbf{C}=c)P(d_q)} \\ &= \log \frac{P(\mathbf{C}=c|d_q)}{P(\mathbf{C}=c)} \\ &= \log P(\mathbf{C}=c|d_q) - \log P(\mathbf{C}=c). \end{aligned} \tag{2.4}$$

As $P(\mathbf{C} = c)$ is a constant prior, the problem boils down to computing the posterior $P(\mathbf{C} = c|d_q)$. To enable an efficient computation, we approximate this probability with a random forest.

2.2.2 Random Forest-Based Voting

Random forest was first proposed to solve the classification problem [2]. Later, it has been extended to handle regression problems and is used for many multimedia applications, like [1, 5, 8–11, 14]. In this section, random forest is used to estimate the posterior probability $P(\mathbf{C} = c|d_q)$.

To build the forest from a training dataset, we use a method motivated by [5]. However, compared with [5], which treats a random forest as a classifier and votes for the hypothesis given a feature point, our random forest is used to estimate the posterior distribution of each STIP point.

Two kinds of descriptors for STIP: Histogram of Gradient (HoG) and Histogram of Flow (HoF), are used to build the random forest. In the following, we first describe how to build a single decision tree, then the forest is constructed by M independent trees. Assume we have N STIP points in the training set, defined as $\{(x_i, y_i), i = 1, 2, \ldots, N\}$, where $x_i = (x_i^1, x_i^2)$; $x_i^1 \in R^{72}$ and $x_i^2 \in R^{90}$ refer to the HOG feature and HOF feature, respectively; $y_i \in C$ is the label of the STIP (if we want to detect actions from category $\mathbf{C} = c$, we consider STIPs with $y_i = c$ as positive examples and other STIPs as negative examples). In order to build a tree and split the training set, a random number $\tau \in \{1, 2\}$ is first generated to indicate which kind of feature to use for splitting ($x_i^{\tau=1}$ refers to HOG feature and $x_i^{\tau=2}$ refers to HOF feature). Then two more random integer numbers e_1 and e_2 will be generated, indicating the dimension indices of either HOG or HOF feature. After that, a "feature difference" can be evaluated with $D_i = x_i^\tau(e_1) - x_i^\tau(e_2), i = 1, 2, \ldots, N$. For each x_i, we assign it to the left child node if $x_i^\tau(e_1) - x_i^\tau(e_2) \geq \theta$ or right child node if $x_i^\tau(e_1) - x_i^\tau(e_2) < \theta$.

The threshold θ is selected by minimizing the binary classification error:

$$\theta^* = \operatorname{argmin}_\theta (\min\{\mathcal{E}(c)^L + \mathcal{E}(\bar{c})^R, \mathcal{E}(c)^R + \mathcal{E}(\bar{c})^L\}), \tag{2.5}$$

where:

$$
\begin{aligned}
\mathcal{E}(c)^L &= \sum_{i=1}^N I(y_i \neq c) I(x_i^\tau(e_1) - x_i^\tau(e_2) \geq \theta) \\
\mathcal{E}(c)^R &= \sum_{i=1}^N I(y_i \neq c) I(x_i^\tau(e_1) - x_i^\tau(e_2) < \theta) \\
\mathcal{E}(\bar{c})^L &= \sum_{i=1}^N I(y_i = c) I(x_i^\tau(e_1) - x_i^\tau(e_2) \geq \theta) \\
\mathcal{E}(\bar{c})^R &= \sum_{i=1}^N I(y_i = c) I(x_i^\tau(e_1) - x_i^\tau(e_2) < \theta).
\end{aligned}
\tag{2.6}
$$

In Eq. 2.6, $I(x)$ is a indicator function, that is $I(x) = 1$ if $x = 1$ and 0 otherwise. And c is the action type we want to detect. The first two terms refer to the misclassification errors of the left node and right node, respectively, when the labels of the nodes are both c. The last two terms refer to the misclassification errors of the left node and right node, respectively, when the labels of the nodes are not c.

The above three parameters (τ, e_1 and e_2) can be integrated into a single hypothesis. For example, we can generate a hypothesis to partition the dataset using the following three steps:

- Generate $\tau \in \{1, 2\}$ to indicate the feature type to use
- Generate the dimension index e_1 and e_2 and compute the feature difference $D_i = x_i^\tau(e_1) - x_i^\tau(e_2), i = 1, 2, \ldots, N$

- Split the dataset into two parts based on a threshold on feature difference and obtain a misclassification error

We generate γ hypotheses independently ($\gamma = 200$ in our experiments) and select the one with the smallest misclassification error. After this, one node will be built and the training set will be partitioned into two parts. For each part, a new node will be further constructed in the same way. This process is repeated until any of the two conditions below is satisfied: (1) the depth of the tree reaches the maximum number or (2) the number of points in the node is smaller than a predefined threshold.

Now, we discuss how to compute $P(\mathbf{C} = c | d_q)$ with a random forest. Suppose we have M trees in a forest and the STIP d_q will fall in one of the leaves in a tree. Assume that for a tree T_i, the STIP point d_q falls in a leaf with N_i^+ positive samples and N_i^- negative samples. The posterior distribution of d_q can be approximated by the average density of the M nodes in M different trees:

$$P(\mathbf{C} = c | d_q) \approx \frac{1}{M} \sum_{i=1}^{M} \frac{N_i^+}{N_i^+ + N_i^-}. \tag{2.7}$$

Then Eq. 2.4 can be replaced with:

$$\begin{aligned} S^c(d_q) &= \log P(\mathbf{C} = c | d_q) - \log P(\mathbf{C} = c) \\ &= \log \frac{1}{M} \sum_{i=1}^{M} \frac{N_i^+}{N_i^+ + N_i^-} - \log P(\mathbf{C} = c). \end{aligned} \tag{2.8}$$

In the training dataset, the numbers of STIP points are different for different action classes. Therefore, it is inaccurate to compute the prior probability $P(\mathbf{C} = c)$ directly from the distribution of training dataset. In our experiments, we introduce the parameter $A = -\log P(\mathbf{C} = c)$ and optimize it in the experiments.

The benefits of using the random forest are numerous. First, each tree in the forest is independent to other trees when evaluating $P(\mathbf{C} = c | d_q)$ in Eq. 2.7. The average of them thus reduces the variance of the estimation. Second, random forest is fast to evaluate during the testing stage. The runtime cost for each STIP only depends on the depth of each tree and the number of trees. It is not affected by the number of points in the training data. Hence, it is much faster than LSH-based nearest neighbor search. In the experiment section, we will show that random forest-based voting approach is over 4,000 times faster than the LSH-based approach. Another advantage of random forest compared with LSH is that, when constructing the trees, the label information of x_i can be integrated. Thus, the trees follow the data distribution of the training data. This improves the generalization ability. Finally, the construction of random forest is flexible. Besides the label information, it is easy to combine other types of feature descriptors and spatial information of STIPs.

According to the literature, [10, 14, 15] also employ tree structures for action recognition. We first consider the differences between [14] and our work. The feature [14] employs is densely sampled while we use the sparse STIP features. Second, [14] votes for the center of the action while our random forest weighs each STIP point so that the non-trivial scale estimation can be partially solved with branch and bound search. Third, the votes in [14] are estimated from the frequency view so that it would generate positive votes even for the background. On the contrary, our votes employs the mutual information-based measure (Eq. 2.4), which is more discriminative thanks to the introduction of negative votes. The trees in [10] are used for indexing and searching nearest neighbor while trees in [15] serves as a codebook. Since we employ random forest to weigh each STIP points, the motivations and implementations are different from [10, 15]. Besides, our work can deal with not only action classification but also action detection, while [10, 15] are only applicable to action recognition.

After obtaining the individual voting score of each STIP, the spatio-temporal location and scale of the target action will be determined by the branch-and-bound search as described in next section.

2.3 Action Detection and Localization

The purpose of action detection is to find a subvolume V with the maximum similarity to the pre-defined action type. Following [16], with each STIP being associated with an individual score $s^c(d)$, our goal is to find the video subvolume with the maximum score:

$$V^* = \text{argmax}_{V \subset \mathcal{V}} f(V), \qquad (2.9)$$

where, $V = [T, B] \times [L, R] \times [S, E]$ is a video subvolume, where L, R, T, B, S and E are the left, right, top, bottom, start and end positions of V; $f(V) = \sum_{d \in V} s^c(d)$ and $\mathcal{V}$ is the whole video space. A subvolume V is said to be *maximal* if there does not exist any other subvolume V' such that $f(V') > f(V)$ and $V' \cap V \neq \emptyset$. The action detection problem is to find all the maximal subvolumes whose scores are above a certain threshold.

A spatio-temporal branch-and-bound algorithm was proposed in [16] to solve the single subvolume search problem. Instead of performing a branch-and-bound search directly in the 6-dimensional parameter space Λ, the method performs a branch-and-bound search in the four-dimensional spatial parameter space. In other words, it finds the spatial window W^* that maximizes the following function:

$$F(W) = \max_{T \subseteq \mathbb{T}} f(W \times T), \qquad (2.10)$$

where $W = [T, B] \times [L, R]$ is the spatial window; $T = [S, E]$ is the temporal segment, and $\mathbb{T} = [0, t - 1]$.

One advantage of separating the parameter space is that the worst case complexity is reduced from $O(m^2 n^2 t^2)$ to $O(m^2 n^2 t)$. The complexity is linear in t, which is usually the largest of the three dimensions. For this reason, it is efficient in processing long videos, but when the spatial resolution of the video increases, the complexity goes up quickly. The method in [16] was tested on videos with low resolution (160×120). In this section, we are interested in higher resolution videos (320×240 or higher). We found that for videos taken under challenging lighting conditions with crowded background such as those in the publicly available MSR Action dataset II,[1] the action detection rates on 320×240 resolution videos are much better than those on 160×120. Unfortunately, the subvolume search for 320×240 videos is much slower. For example, [16] takes 20 h to search the MSR Action dataset II which consists of 54 video sequences of 1 min long each with 320×240 resolution.

Moreover, in [16], the multi-instance detection problem was converted to a series of single subvolume search problem. They first find the optimal subvolume V_1 such that $f(V_1) = max_V f(V)$. After that, it sets the scores of all the points in V_1 to 0, and finds the optimal subvolume V_2, and so on. To further speed up the search process during the branch-and-bound iterations, a heuristic was used in [18]. If a candidate window $\mathbb{W}$ with a score larger than the detection threshold is found, the subsequent searches are limited to the subwindows contained in $\mathbb{W}$. It guarantees that it will find a valid detection, but the detected subvolume is not guaranteed to be optimal.

In the next two subsections, we present two techniques to speed up the subvolume search algorithm. The combination of the two techniques allows us to perform subvolume search on 320×240 videos in real time.

2.3.1 Spatial Down-Sampling

To handle high-resolution videos, the technique is to spatially down-sample the video space by a factor s before the branch-and-bound search. Note that the interest point detection, descriptor extraction, and the scores are all done in the original video sequence.

For a video volume V of size $m \times n \times t$, the size of the down-sampled volume V^s with scale factor s is $\frac{m}{s} \times \frac{n}{s} \times t$. For any point $(i, j, k) \in V^s$ where $i \in [0, \frac{m}{s} - 1]$, $j \in [\frac{n}{s} - 1]$, and $k \in [0, t - 1]$, its score is defined as the sum of the scores of the $s \times s$ points in V, that is, $f^s(i, j, k)$ is defined as

$$f^s(i, j, k) = \sum_{x=0}^{s-1} \sum_{y=0}^{s-1} f(s * i + x, s * j + y, k). \qquad (2.11)$$

Given any subvolume $V^s = [L, R] \times [T, B] \times [S, E] \subset V^s$, where L, R, T, B, S and E are the left, right, top, bottom, start and end positions of V^s, respectively, denote $\xi(V^s)$ to be its corresponding subvolume in original video V, that is,

$$\xi(V^s) = [s * L, s * (R + 1) - 1] \times [s * T, s * (B + 1) - 1] \times [S, E]. \qquad (2.12)$$

---

[1] The MSR action dataset II is available at http://research.microsoft.com/en-us/um/people/zliu/ ActionRecoRsrc/default.htm.

As they are the same subvolume, it is easy to see that

$$f^s(V^s) = f(\xi(V^s)). \tag{2.13}$$

A subvolume $V = [X_1, X_2] \times [Y_1, Y_2] \times [T_1, T_2] \subset \mathcal{V}$ is called an *s-aligned* subvolume if X_1 and Y_1 are multiples of s and the width $X_2 - X_1 + 1$ and height $Y_2 - Y_1 + 1$ are also multiples of s. Equation 2.12 provides a one-to-one mapping between the volumes in $\mathcal{V}^s$ and the s-aligned subvolumes in $\mathcal{V}$.

Instead of searching the original video space, we can search the down-sampled video space $\mathcal{V}^s$ of a much smaller size $\frac{m}{s} \times \frac{n}{s} \times t$. However, as the down-sampling process also introduces the approximation errors, it affects the search results. In general, for any $V^s \subset \mathcal{V}^s$, there exists a $V = \xi(V^s) \subset \mathcal{V}$. It thus shows that the maximum subvolume found in the down-sampled space is at most as good as the one found in the original space:

$$\max{}_{V^s \subset \mathcal{V}^s} f^s(V^s) \leq \max{}_{V \subset \mathcal{V}} f(V). \tag{2.14}$$

We illustrate a concrete example in Fig. 2.2. For simplicity, in Fig. 2.2, we choose the down-sampling factor $s = 2$ and discuss the problem in the $2D$ space (only one frame is considered). The left figure shows the original video space and its down-sampled version is in the right figure. Each pixel is associated with a voting score. The orange rectangle highlights the optimal solution in the original video space, namely the bounding box of the highest total sum. After the down-sampling, the gray rectangle is the detection result in the down-sampled video. By mapping it back to the original space, we obtain an approximate solution highlighted by the red rectangle. It overlaps with the optimal solution in the original space, but the total sum is slightly less. To further quantify the approximation error, we derive the upper bound of the error caused by the down-sampling, as explained in Theorem 2.1.

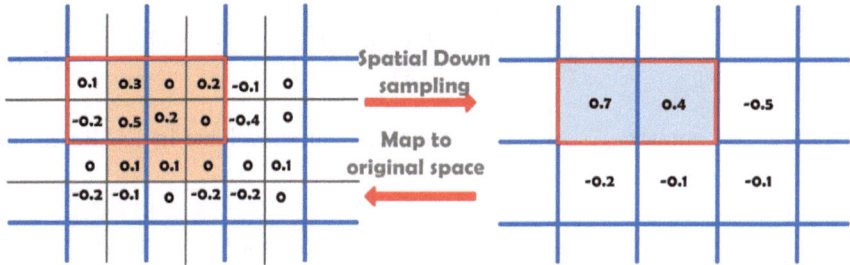

Fig. 2.2 Approximation of the spatial down-sampling. *Left* figure shows the score image in the original resolution and *right* figure shows the down-sampled score image. Every four small pixels in a cell from the original resolution sum up to one score in the low resolution, for example, the value in the *top-left* pixel from the *right* figure $0.5 = 0.1 + 0.3 - 0.4 + 0.5$. We notice that the optimal solution found in the down-sampled video space is worse than that in the original space ($f^s(\tilde{V}^*) = 1.1 < f(V^*) = 1.4$)

Theorem 2.1 *Bound of the approximation error. Let* V^* *denote the optimal subvolume in* $\mathcal{V}$*, that is,* $f(V^*) = max_{V \subset \mathcal{V}} f(V)$*. Assume* $V^* = [x_1, x_1 + w - 1] \times [y_1, y_1 + h - 1] \times [t_1, t_2]$ *where w and h are the width and height of* V^**, respectively and further assume the total score of a subvolume is on average proportional to its size. Then, there exists an s-aligned subvolume* $\tilde{V}$ *satisfying:*

$$f(\tilde{V}) \geq (1 - \frac{s*h + s*w + s^2}{wh})f(V^*). \tag{2.15}$$

Proof Let V^* denote the optimal subvolume in $\mathcal{V}$, that is, $f(V^*) = max_{V \subset \mathcal{V}} f(V)$. Assume $V^* = [x_1, x_1 + w - 1] \times [y_1, y_1 + h - 1] \times [t_1, t_2]$ where w and h are the width and height of V^*, respectively. Let $|V|$ denote the number of voxels in V. It can be shown that there exists an s-aligned subvolume $\tilde{V} = [\tilde{x}_1, \tilde{x}_1 + \tilde{w} - 1] \times [\tilde{y}_1, \tilde{y}_1 + \tilde{h} - 1] \times [t_1, t_2]$ such that

$$|(V^* \setminus \tilde{V}) \cup (\tilde{V} \setminus V^*)| \leq (s*h + s*w + s^2)(t_2 - t_1). \tag{2.16}$$

Therefore,

$$\frac{|(V^* \setminus \tilde{V}) \cup (\tilde{V} \setminus V^*)|}{|V^*|} \leq \frac{s*h + s*w + s^2}{wh}. \tag{2.17}$$

If we assume the total score of a subvolume is on average proportional to its size, then

$$\frac{f((V^* \setminus \tilde{V}) \cup (\tilde{V} \setminus V^*))}{f(V^*)} \leq \frac{s*h + s*w + s^2}{wh}. \tag{2.18}$$

Therefore

$$\frac{f(V^*) - f(\tilde{V})}{f(V^*)} \leq \frac{s*h + s*w + s^2}{wh}. \tag{2.19}$$

After a re-arrangement of the items, we have:

$$f(\tilde{V}) \geq (1 - \frac{s*h + s*w + s^2}{wh})f(V^*). \tag{2.20}$$

Let $\tilde{V}^* = argmax_{V \in \mathcal{V}^s} f^s(V)$ denote the optimal subvolume in $\mathcal{V}^s$. Based on Eq. 2.15, we have

$$f^s(\tilde{V}^*) \geq (1 - \frac{s*h + s*w + s^2}{wh})f(V^*). \tag{2.21}$$

As an example, suppose the spatial dimension of V is 320×240, and the scale factor $s = 8$. The spatial dimension of the down-sampled volume is 40×30. If we assume the window size of the optimal subvolume V^* is 64×64, then the average relative error is at most

$$\frac{s*h+s*w+s^2}{wh}=\frac{8*64+8*64+8^2}{64^2}\approx 25\,\%. \tag{2.22}$$

We have run numerical experiments to measure the relative error of the optimal solutions in the down-sampled volumes. We used 30 video sequences of resolution 320×240. There are three action types. For each video sequence and each action type, we obtain a 3D volume of scores as defined in Eq. 2.8. We choose $s = 8$, and down-sample each 3D volume to spatial resolution of 40×30. There are 113 actions in total. For each action, we compute its corresponding down-sampled subvolume, and evaluate the relative error which is the score difference divided by the original action score. The mean is 23 % and the standard deviation is 26 %. We can see that the numerical experiments are consistent with the theoretical analysis.

2.3.2 Top-K Search Algorithm

The multi-instance search algorithm in [16] repeatedly applies the single-instance algorithm many times until some stop criteria is met. In practice, there are typically two different stop conditions that can be used. The first is to stop after k iterations where k is a user-specified integer. The second is to stop when the detection score is smaller than a user-specified detection threshold λ. In either case, suppose the number of detected instances is k, then the worst case complexity of the algorithm is $O(kn^2m^2t)$.

We notice that in 1D case, Brodal and Jorgensen [3] developed an algorithm that finds the Top-K subarrays in $O(n + k)$ time. This is much more efficient than repeatedly applying the single-instance algorithm k times which has the complexity $O(kn)$. In 3D case, we would also like to have an algorithm that is more efficient than simply applying the single-instance algorithm k times. We consider two different variants corresponding to the two stop criteria. The first, called λ search, can be applied when we are interested in finding all the subvolumes above a user-specified threshold λ. The second, called Top-K search, can be applied when we are interested in finding the Top-K subvolumes.

2.3.2.1 λ Search

In this section, we describe an algorithm that finds all of the subvolumes with scores larger than a user-specified threshold λ. The pseudo-code of the algorithm is shown in Algorithm 1. Following the notation in [16], we use $\mathbb{W}$ to denote a collection of spatial windows, which is defined by 4 intervals that specify the parameter ranges for the left, right, top, and bottom positions, respectively. Given any set of windows $\mathbb{W}$, we use $\hat{F}(\mathbb{W})$ to denote its upper bound which is estimated in the same way as in [16, 17]. We use $W_{\max}$ to denote the largest window among all the windows in $\mathbb{W}$. Initially, $\mathbb{W}^*$ is equal to the set of all the possible windows on the image and $F*$ is the corresponding upper bound, as in Line 5 of Algorithm 1. From Line 6–19, we split and store the results if the top state $\mathbb{W}$ is over a threshold λ and iterate this process. From Line 20–22, we have a subvolume (V^*) detected. The whole process iterates until the score for the detected subvolume is below the threshold.

Algorithm 1 λ search.

1: Initialize P as empty priority queue
2: set $\mathbb{W} = [T, B, L, R] = [0, m] \times [0, m] \times [0, n] \times [0, n]$
3: push($\mathbb{W}, \hat{F}(\mathbb{W})$) into P
4: **repeat**
5:   Initialize current best solution F^*, W^*
6:   **repeat**
7:     retrieve top state $\mathbb{W}$ from P based on $\hat{F}(\mathbb{W})$
8:     **if** $\hat{F}(\mathbb{W}) > \lambda$ **then**
9:       split $\mathbb{W}$ into $\mathbb{W}^1 \cup \mathbb{W}^2$
10:       **if** $\hat{F}(\mathbb{W}^1) > \lambda$ **then**
11:         push ($\mathbb{W}^1, \hat{F}(\mathbb{W}^1)$) into P
12:         update current best solution $\{W^*, F^*\}$
13:       **end if**
14:       **if** $\hat{F}(\mathbb{W}^2) > \lambda$ **then**
15:         push ($\mathbb{W}^2, \hat{F}(\mathbb{W}^2)$) into P
16:         update current best solution $\{W^*, F^*\}$
17:       **end if**
18:     **end if**
19:   **until** $\hat{F}(\mathbb{W}) \leq F^*$
20:   $T^* = argmax_{T \in [0,t]} f(W^*, T)$;
21:   add $V^* = [W^*, T^*]$ to the list of detected subvolumes.
22:   for each point $(i, j, k) \in V^*$, set $f(i, j, k) = 0$.
23: **until** $\hat{F}(\mathbb{W}) \leq \lambda$

In terms of the worst case complexity, the number of branches of this algorithm is no larger than $O(n^2 m^2)$ since the algorithm does not re-start the priority queue P. Each time it branches, the algorithm has to compute the upper bound whose complexity is $O(t)$. Therefore the worst complexity involved in branch and bound is the same as [16]: $O(n^2 m^2 t)$. In addition, each time when it detects a subvolume, the algorithm has to update the scores of the video volume which has complexity $O(nmt)$. If there are k detected subvolumes, the complexity for updating the scores is $O(kmnt)$. Overall, the worst case complexity of this algorithm is $O(n^2 m^2 t) + O(kmnt)$. When k is large, this is much better than $O(kn^2 m^2 t)$.

2.3.2.2 Top-K Search

In this section, we describe how to modify Algorithm 2 for the case when we are interested in finding the Top-K actions, and we assume we do not know the threshold λ.

The pseudo-code of the algorithm is shown in Algorithm 2. The algorithm is similar to Algorithm 2. In Line 6, $(\{W_i^*, F_i^*\})_{i=c...k}$ are set as all the possible windows on the image and its upper bound score, respectively. From Line 6–20, we split and store the results if the top state $\mathbb{W}$ is over the Kth top score and iterate this process. From Line 21–24, we have a subvolume (V^*) detected. The whole process iterates until K subvolumes are detected. There are four major differences. First, instead of maintaining a single current best solution, it maintains k-best current solutions. Second, it replaces the criteria $\hat{F}(\mathbb{W}) > \lambda$ with $\hat{F}(\mathbb{W}) > F_k^*$ to determine whether we need to insert $\mathbb{W}^1$ or $\mathbb{W}^2$ into the queue P. Third, it replaces the inner-loop stop criteria $\hat{F}(\mathbb{W}) \leq F^*$ with $\hat{F}(\mathbb{W}) \leq F_c^*$. Finally, the outer-loop stop criteria $\hat{F}(\mathbb{W}) \leq \lambda$ is replaced with $c > k$. In this algorithm, the number of outer loops is k. So the worst case complexity is also $O(n^2m^2t) + O(kmnt)$.

Algorithm 2 Top-K Search.

1: Initialize P as empty priority queue
2: set $\mathbb{W} = [T, B, L, R] = [0, m] \times [0, m] \times [0, n] \times [0, n]$
3: push($\mathbb{W}, \hat{F}(\mathbb{W})$) into P
4: c=1
5: **repeat**
6:    Initialize $(\{W_i^*, F_i^*\})_{i=c...k}$ where $F_k^* \leq ... \leq F_c^*$
7:    **repeat**
8:       retrieve top state $\mathbb{W}$ from P based on $\hat{F}(\mathbb{W})$
9:       **if** $\hat{F}(\mathbb{W}) > F_k^*$ **then**
10:          split $\mathbb{W}$ into $\mathbb{W}^1 \cup \mathbb{W}^2$
11:          **if** $\hat{F}(\mathbb{W}^1) > F_k^*$ **then**
12:             push $(\mathbb{W}^1, \hat{F}(\mathbb{W}^1))$ into P
13:             update $(\{W_i^*, F_i^*\})_{i=c...k}$
14:          **end if**
15:          **if** $\hat{F}(\mathbb{W}^2) > F_k^*$ **then**
16:             push $(\mathbb{W}^2, \hat{F}(\mathbb{W}^2))$ into P
17:             update $(\{W_i^*, F_i^*\})_{i=c...k}$
18:          **end if**
19:       **end if**
20:    **until** $\hat{F}(\mathbb{W}) \leq F_c^*$
21:    $T^* = argmax_{T \in [0,t]} f(W^*, T)$;
22:    output $V_c^* = [W^*, T^*]$ as the c-th detected subvolume
23:    for each point $(i, j, k) \in V_c^*$, set $f(i, j, k) = 0$.
24:    c = c+1
25: **until** $c > k$

2.4 Experiments

2.4.1 Action Classification

To evaluate our proposed random forest-based approach for multiclass action classification, we test on the benchmark KTH dataset. The experiment setup is the same as [7, 16] where clips from 16 persons are used for training and the other 9 persons

Table 2.1 Confusion matrix for KTH action dataset

	Clap	Wave	Box	Run	Jog	Walk
Clap	137	1	6	0	0	0
Wave	7	137	0	0	0	0
Box	0	0	144	0	0	0
Run	0	0	0	95	47	2
Jog	0	0	0	4	136	4
Walk	0	0	0	0	0	144

The total accuracy is 91.8 %

Table 2.2 Comparison of different reported results on KTH dataset

Method	Mean accuracy (%)
Our method	91.8
Yuan et al.'s [16]	93.3
Reddy et al.'s [10]	90.3
Laptev et al.'s [6]	91.8

are used for testing. The confusion matrix is listed in Table 2.1. We also compare our results with the state-of-the-art results in Table 2.2. With the same input features, our method performs as well as the method using support vector machine for classification [6]. Although our performance is slightly worse than the nearest neighbor-based classification in [16], as will be shown later, our approach is significantly faster as it avoids the nearest neighbor search.

2.4.2 Action Detection

To evaluate our multiclass action detection and localization, we perform cross-dataset training and testing. We first build a random forest using the KTH dataset (with the 16 persons in the training part) and then test on a challenging dataset (MSRII) of 54 video sequences where each video consists of several actions performed by different people in a crowded environment. Each video is approximately one minute long. The videos contain three different types of actions: handwaving, handclapping, and boxing. Some videos contain different people performing different actions simultaneously. There are also instances where a person performs two different actions consecutively.

For all of our experiments we have fixed $K = 3$, $\lambda = 3.0$. Moreover, unless explicitly mentioned we down-sample the score volume to 40×30 pixels.

Figure 2.3 compares the precision-recall for the following methods (the original videos are of high resolution 320×240):

(i) ASTBB (Accelerated Spatio-Temporal Branch-and-Bound search) of [18] in low resolution score volume (frame size 40×30),
(ii) ASTBB of [18] in 320×240 videos,

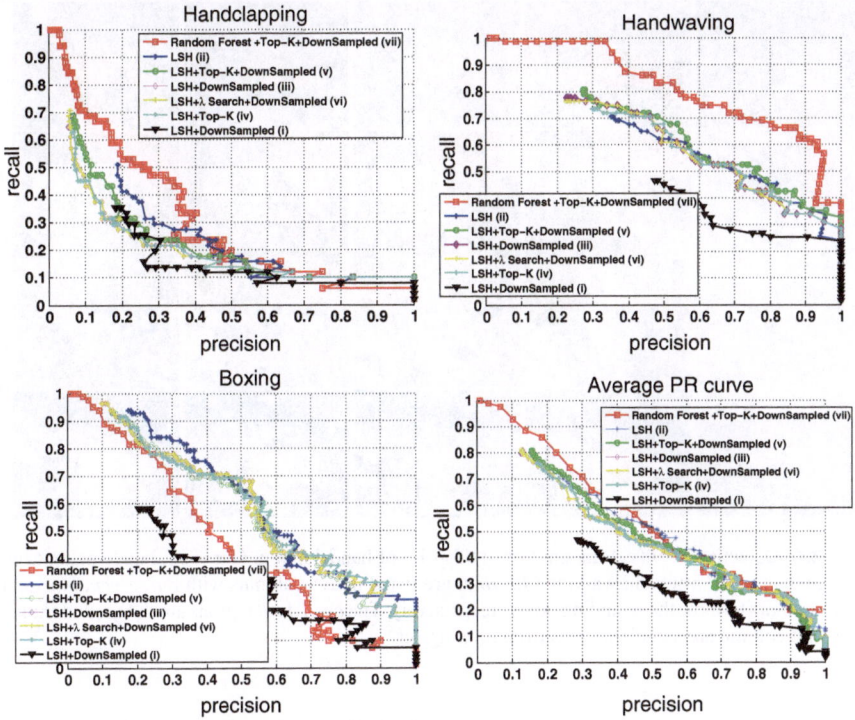

Fig. 2.3 Precision-recall curves for action detections with different methods

(iii) multiround branch-and-bound search of [16] in low-resolution score volume (frame size 40 × 30),

(iv) Top-K search at original size 320 × 240,   '

(v) Top-K search at down-sampled score volume (size 40 × 30),

(vi) λ search at down-sampled score volume (size 40 × 30),

(vii) random forest-based weighting followed by Top-K search at down-sampled score volume (size 40 × 30).

Except for (vii), which uses our random forest-based voting score, the other methods apply the LSH-based nearest neighbor voting score as in [16]. The parameter $A = -\log P(C = c)$ in Eq. 2.8 for method (vii) is set to 2.1, 1.7 and 0.9 for handclapping, handwaving and boxing respectively. Also, we use the walking actions from KTH as the negative dataset when constructing forests. For the purpose of generating precision-recall curves, we modified the outer-loop stop criteria (Line 25, Algorithm 2) to repeat until $\hat{F}(\mathbb{W}) \leq \lambda$ where λ is a small threshold. In this way, it outputs more than K subvolumes which is necessary for plotting the precision-recall curve. Some sample detection results obtained by our approach (vii) are shown in Fig. 2.4. To demonstrate the capability of handling non-stationary actions, we show a walking detection result at the bottom row. The detection is done by using KTH

Fig. 2.4 Detection results (Random Forest+Top-K) of handclapping (*1st row*), handwaving (*2nd row*), boxing (*3rd row*) and walking (*4th row*) are listed in 2-5 columns with *red, green, blue* and *yellow* colors to show the bounding boxes, respectively. The *cyan dash* regions are the ground truths. The first column are sample images from training set

walking as the positive training data while the KTH handwaving, handclapping, and boxing are used as the negative training data.

The measurement of precision and recall is the same as what is described in [16]. For the computation of the precision we consider a true detection if : $\frac{\text{Volume}(V^* \cap G)}{\text{Volume}(G)} > \frac{1}{8}$ where G is the annotated ground truth subvolume, and V^* is the detected subvolume. On the other side, for the computation of the recall we consider a hit if: $\frac{\text{Volume}(V^* \cap G)}{\text{Volume}(V^*)} > \frac{1}{8}$.

We first compare the results based on LSH voting approaches. Figure 2.3 lists the Precision-Recall curves for the three different action classes, respectively. The average PR curve is computed by averaging precision and recall results among the three action types while adjusting a threshold. This can give a general idea of the overall performance for different algorithms. From the precision-recall curves, we can see that although the accelerated search of [18] provides excellent results in high resolution videos, its performance on down-sampled low resolution videos is poor compared with other search schemes. Moreover, all the methods applied to the high resolution videos provide similar performance. In particular, the methods of Top-K search with branch-and-bound search at down-sampled size (v) and λ search with branch-and-bound search at down-sampled size (vi) are among the best ones. These results justify our proposed λ search and Top-K search algorithms. Although the branch-and-bound is performed in the down-sampled size videos, it still provides good performance. However, the search speed is much faster. To compare the performance of action detection between LSH and random forest, (v) and (vii) are two search schemes with the same environment but different voting approaches. Random forest (vii) is superior to LSH (v) in handwaving but poorer in boxing. Since the boxing action is highly biased in KTH dataset (much more boxing actions are performed from right to left), it reduces the discriminative ability of the trees. For LSH, however, because it searches only one nearest positive and negative in the neighborhood, the effect of such bias can almost be ignored.

2.4.3 Computational Cost

The feature extraction step is performed with publicly available code in [7]. Although their code may not be very fast, there are faster implementations available. Therefore, the computation time for feature extraction is not considered in this section. We suppose all the STIP points are already extracted and stored in the memory. Then, the computational time of our algorithms is dominated by two operations, computing the score for each STIP and branch-and-bound search. For the first part, LSH takes on average 18.667 ms per STIP point while random forest only takes 0.0042 ms. To deal with a video clip with 10,000 STIPs, it will take around 186.67 s for LSH but only 42 ms for random forest. That is, random forest-based approach is 4,000 times faster than LSH-based approach.

Table 2.4 shows the time consumed for the search part. All of the algorithms are implemented using C++, performed on a single PC of dual-core and 4G main memory: (a) accelerated λ search of [18] in low resolution videos (frame size 40 $\times$ 30), (b) accelerated λ search of [18] in high resolution videos, (c) multiround branch-and-bound search of [16] in low-resolution videos (frame size 40 $\times$ 30), (d) λ search, with branch-and-bound search at down-sampled size 40 $\times$ 30, (e) Top-K search, with branch-and-bound search at down-sampled size 80 $\times$ 60, (f) Top-K search, with branch-and-bound search at down-sampled size 40 $\times$ 30.

Table 2.4 shows that although the method of [18] works well for low resolution videos, the search speed becomes much slower for high resolution videos. Moreover as shown in Fig. 2.3, when performing on the down-sampled score volumes, the heuristic method of [18] (curve (i)) is a lot worse than the other methods. This is an indication that it is not a good idea to perform heuristic search on down-sampled score volumes. In comparison, λ search provides much better search quality. Among all the search schemes, the fastest method is the Top-K search with branch-and-bound

Table 2.3 Time consumed for voting one STIP and one video sequence (for example, 10,000 STIP points)

Method	Voting time (ms)	One sequence (s)
LSH	18.667 ± 8.4105	186.67
Random forest	0.0042 ± 0.0032	0.042

Only CPU time is considered

Table 2.4 Time consumed for each method to search actions in the 54 videos

Method	Running time
Low resolution (40 × 30) [18]	40 min
High resolution (320 × 240) [18]	20 h
Down-sampled B&B (40 × 30)	10 h
λ search + down-sampled B&B (40 × 30)	1 h 20 min
Top-K + down-sampled B&B (80 × 60)	6 h
Top-K + down-sampled B&B (40 × 30)	26 min

Table 2.5 Comparison of total time cost for action detection

Method	Voting time (mins)	Search time (min)	Total time (min)
LSH+B&B [18]	271	1,200	1,471
LSH+Top-K (our algorithm)	271	26	297
Random forest+Top-K (our algorithm)	0.62	26	26.62

Only CPU time is considered

at down-sampled score volume of 40 × 30. It takes only 26 min to process the 54 sequences whose total length is about one hour in total.

Finally, we compare LSH and random forest in terms of total computation time in Table 2.5, including the runtime cost for computing scores and the runtime cost for top-K search. For the previous method [18], it takes at least 1,471 min to search all the actions for 54 videos in MSRII. In contrast, the total computation time of our proposed algorithm is 26.62 min.

2.5 Summary of this Chapter

We have developed a novel system for human action recognition and spatio-temporal localization in video sequences. The system improves upon the state-of-arts on two aspects. First, we proposed a random forest-based voting technique to compute the scores of the interest points, which achieves multiple orders-of-magnitude speed-up compared to the nearest neighbor-based scoring scheme. Second, we proposed a top-k search technique which detects multiple action instances simultaneously with a single round of branch-and-bound search. To reduce the computational complexity of searching higher resolution videos, we performed subvolume search on the down-sampled score volumes. We have presented experimental results on challenging videos with crowded background. The results showed that our proposed system is robust to dynamic and cluttered background and is able to perform efficient action detection on high resolution videos. In next chapter, we will further improve the computational speed for action localization with a coarse-to-fine branch and bound search strategy, and extend the idea for the human action search based on one query sample.

References

1. A. Bosch, A. Zisserman, X. Munoz, Image classification using random forests and ferns, in *Proceedings of the IEEE International Conference on Computer Vision* (2007)
2. L. Breiman, Random forests. Mach. Learn. **45**, 5–32 (2001)
3. G. Brodal, A. Jørgensen, A linear time algorithm for the k maximal sums problem. Math. Found. Comput. Sci. **4707**, 442–453 (2007)
4. K.G. Derpanis, M. Sizintsev, K. Cannons, R.P. Wildes, Efficient action spotting based on a spacetime oriented structure representation, in *Computer Vision and Pattern Recognition (CVPR)* (2010)
5. J. Gall, V. Lempitsky, Class-specific hough forests for object detection, in *Proceedings of the IEEE Conference on Computer Vision and Pattern Recognition* (2009)
6. I. Laptev, M. Marszalek, C. Schmid, B. Rozenfeld, Learning realistic human actions from movies, in *Proceedings of the IEEE Conference on Computer Vision and Pattern Recognition* (2008)
7. I. Laptev, On space-time interest points. Int. J. Comput. Vis. **64**(2–3), 107–123 (2005)
8. V. Lepetit, P. Lagger, P. Fua, Randomized trees for real-time keypoint recognition, in *Computer Vision and Pattern Recognition*(textitCVPR) (2005)
9. K. Mikolajczyk, H. Uemura, Action recognition with motion-appearance vocabulary forest, in *Computer Vision and Pattern Recognition (CVPR)* (2008)
10. K.K. Reddy, J. Liu, M. Shah, Incremental action recognition using feature-tree, in *Proceedings of the IEEE International Conference on Computer Vision* (2009)
11. F. Schroff, A. Criminisi, A. Zisserman, Object class segmentation using random forests, in *Proceedings of the British Machine Vision Conference* (2008)
12. H.J. Seo, P. Milanfar, Action recognition from one example, in *IEEE Transactions on Pattern Analysis and Machine Intelligence (PAMI)* (2010)
13. H. Wang, M.M. Ullah, A. Klaser, I. Laptev, C. Schmid, Evaluation of local spatio-temporal features for action recognition, in *Proceedings of the British Machine Vision Conference* (2009)
14. A. Yao, J. Gall, L. Van Gool, A hough transform-based voting framework for action recognition, in *Computer Vistion and Pattern Recognition (CVPR)* (2010)
15. T.H. Yu, T.K. Kim, R. Cipolla, Real-time action recognition by spatitemporal sematic and structural forest, in *BMVC* (2010)
16. J. Yuan, Z. Liu, Y. Wu, Discriminative subvolume search for efficient action detection, in *Proceedings of the IEEE Conference on Computer Vision and Pattern Recognition* (2009)
17. J. Yuan, Z. Liu, Y. Wu, Discriminative video pattern search for efficient action detection, in *IEEE Transactions on Pattern Analysis and Machine Intelligence (PAMI)* (in press)
18. J. Yuan, Z. Liu, Y. Wu, Z. Zhang, Speeding up spatio-temporal sliding-window search for efficient event detection in crowded videos, in *ACM Multimeida Workshop on Events in Multimedia* (2009)

Chapter 3
Unsupervised Trees for Human Action Search

Abstract Action search is an interesting problem for human action analysis, which
has a lot of potential applications in industry. In this chapter, we propose a very fast
action retrieval system which can effectively locate the subvolumes similar to the
query video. Random-indexing-trees-based visual vocabularies are introduced for
the database indexing. By increasing the number of vocabularies, the large intra-
class variance problem can be relieved despite only one query sample available. In
addition, we use a mutual information based formulation, which is easy to leverage feedback from the user. Also, a coarse-to-fine subvolume search scheme is proposed, which results in a dramatic speedup over the existing video branch-and-bound
method. Cross-dataset experiments demonstrate that our proposed method is not only
fast to search higher-resolution videos, but also robust to action variations, partial
occlusions, and cluttered and dynamic backgrounds. Besides from the superior performance, our system is fast for on-line applications, for example, we can finish an
action search in 24 s from a 1 h database and in 37 s from a 5 h database.

Keywords Action search · 3D Branch-and-bound search · Coarse-to-fine search ·
Multiple visual vocabularies · Random indexing trees

3.1 Introduction

Nowadays, despite text-based search techniques (e.g., Google and Bing search) are
powerful, they can only be applied to searching documents and text-annotated records
with appreciable success and satisfaction achieved. For analyzing and searching
video data, however, such a keyword-based video search is far from satisfactory, and
oftentimes, could yield not-so-useful results. For example, since high-level semantic
meaning is fairly difficult to be described by a limited set of keywords, keyword
based video search is not suitable for detecting a specific human action or video
event, e.g., detecting a pick-up action or a vehicle committing a hit-and-run crime.

Although Chap. 2 proposed an efficient algorithm for human action recognition
and detection, it is still far enough satisfactory for human action search, where a
real-time interaction system is demanded. Besides, for the algorithm in Chap. 2, a

© The Author(s) 2015   29
G. Yu et al., *Human Action Analysis with Randomized Trees*,
SpringerBriefs in Signal Processing, DOI 10.1007/978-981-287-167-1_3

set of positive and negative training videos are required for training the random forest, which is not possible for the human action search with only one query sample provided. Thus, in this chapter, we will devise a novel human action search algorithm, which significantly improve the computational cost for action localization. Besides, feedback from the user is also possible to further improve the retrieval results.

Formally, human action search problem is defined as: given an action example as a query, e.g., hand waving or picking up, the goal is to detect and accurately locate similar actions in the video database. There are three challenges to address for the human action search problem.

First of all, for action search, usually only a single query example is provided. In such a case, the amount of training data is extremely limited and only available at the time of query, whereas in action classification [1, 16, 20] and detection [3, 10, 21], a lot of positive and negative training examples can be leveraged. Therefore, it is much more difficult to identify and locate a specific action example in videos. Furthermore, possible action variations such as scale changes, style changes and partial occlusions only worsen the problem, let alone cluttered and dynamic backgrounds.

Second, a video search engine must have a fast response time because otherwise the user experience would suffer. Unlike video event recognition [5], where the goal is to classify or rank pre-segmented video shots, action search is more difficult as we need to not only recognize the target action, but also locate it accurately, i.e., identify the spatio-temporal extent of the action in the video space. The accurate localization is of great importance especially for the crowded scenes where there are multiple people or moving objects. However, because actions can be small video objects, it is time consuming to locate them in the large video space. In general, for a dataset consisting of tens of hours of videos, such an action search process is expected to finish in just a few seconds.

Finally, a retrieval process typically prefers to enable user interactions, which allows the user to clarify and update their preferences. Thus, a practical action search system must have the flexibility to refine the retrieval results by leveraging the labels resulting from subsequent user feedback. Although relevance feedback is popular in image search, there is much less work that supports interactive action search.

To address the above challenges, we develop an action search system that addresses two key challenges in content based video search: (1) video indexing and (2) action searching. Each video is characterized by a collection of spatio-temporal interest points (STIPs) [13]. To enable fast matching of STIPs between the query action and the video dataset, effective Indexing is required. Bag-of-words (BoW) is a popular solution to index these interest points by clustering them hierarchically [16]. However, as there is only a single vocabulary, it provides one fixed way to quantize the feature space, and inevitably introduces quantization errors. Regardless of how the quantization is done, it typically results in loss of information. Instead of trying to find the best vocabulary as in some previous work [12, 16], we propose to use an ensemble of vocabularies to index the database. Multiple vocabularies provide multiple representations to the data. Thus, the variance of estimation can be reduced as we increase the number of vocabularies.

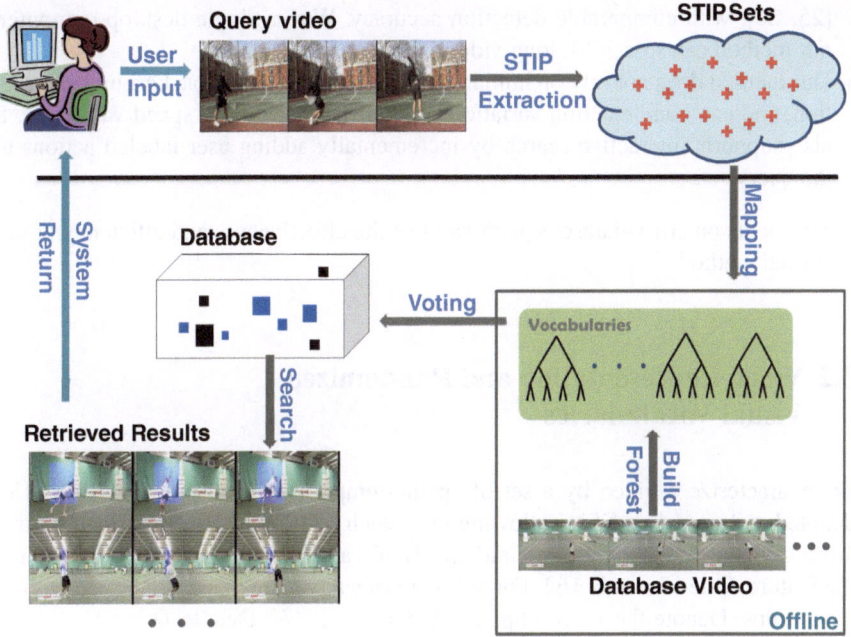

Fig. 3.1 Overview of our algorithm

The second component of our search system is the action search part, i.e., spatially and temporally localize the retrieved actions from the database. Our action search method is based on our previous work of spatio-temporal branch-and-bound search [25, 26] but we further significantly improve the localization speed by introducing a coarse-to-fine search scheme, which is several orders faster than [22, 23, 26].

An overview of our action search system is depicted in Fig. 3.1. The spatial-temporal interest points in videos are first extracted and then labeled according to the exemplar action in the query phase by our random indexing trees. The database can be considered as one large volume, with different values at the positions of the STIP points. The spatio-temporal video subvolumes of highest matching scores, i.e., the summation of scores from all of its interest points, are cropped out as detections. The subvolume is the retrieved result by our algorithm. We summarize the three contributions of our work below.

- We propose to index video interest point features using randomized visual vocabularies to compensate for information loss when using a single visual vocabulary. To implement the randomized visual vocabularies, we use random indexing trees, which provide fast indexing and leads to superior search results.
- For action localization, our proposed coarse-to-fine subvolume search strategy significantly improves the efficiency of the state-of-the-art action detection methods

[25, 26], with comparable detection accuracy. With a single desktop computer, our method can search 5 h long video within only 37.6 s.
- Our method does not rely on human detection, tracking, and background subtraction, and can handle action variations due to small scale and speed variations. It also supports interactive search by incrementally adding user labeled actions to the query set.

Experiments on cross-dataset search validate the effectiveness and efficiency of our proposed method.

3.2 Video Representation and Randomized Visual Vocabularies

We characterize a video by a set of spatial-temporal interest points (STIP) [13], denoted as $V = \{d_i \in \mathbb{R}^n\}$. Following [13], each STIP point d is described by two kinds of features: Histogram of Gradient (HOG) and Histogram of Flow (HOF) and the feature dimension n is 162. For action retrieval, we are given a database with N video clips. Denote the video clips as $\mathcal{V}_i, i = 1, \ldots, N$. Denote $\mathcal{D} = \{\mathcal{V}_1 \cup \mathcal{V}_2 \cup \cdots \cup \mathcal{V}_N\}$. These video clips contain various types of actions such as handwaving, boxing, and walking and last for several hours long.

In order to search for human actions in a large database, indexing becomes one of the most crucial parts. Traditionally, bag-of-words models [12, 16] with hierarchical K-means is widely used for interest points indexing. However, there is usually only a single vocabulary for BoW. The vocabulary quantizes the data in one fixed way. Quantization error would be introduced regardless of how we quantize the data. Intuitively, one vocabulary can be considered as one way of "viewing" the data. Instead of trying to find the best vocabulary as in the pervious work, we propose to use an ensemble of vocabularies. Multiple vocabularies can provide different viewpoints to the data. This can help to increase our performance as we increase the number of vocabularies.

To implement the vocabularies efficiently, we try to use the tree structures [24]. Although there have been a lot of works on the tree structures for computer vision and machine learning applications [2], little work has been done for efficient index with tree structures. KD-tree allows exact NN search but it is inefficient in high dimension cases and only slightly better than the linear search. Hierarchical K-means usually leads to unbalance trees and training is a time-consuming process. To overcome the above problems, we propose the random indexing trees, which can explore the data distribution in the high dimension cases and index the database in an efficient and effective way.

Assume we have N_D STIP points in the dataset, denoted as $\{x_i = (x_i^1, x_i^2), i = 1, 2, \ldots, N_D\}$; $x_i^1 \in \mathbb{R}^{72}$ and $x_i^2 \in \mathbb{R}^{90}$ are the HOG feature and HOF feature, respectively. In order to build a tree and split the dataset, a random number $\tau \in \{1, 2\}$ is first generated to indicate which kind of feature to use for splitting ($x_i^{\tau=1}$ refers to

HOG feature and $x_i^{\tau=2}$ means HOF feature.) Then two more random numbers e_1 and e_2 will be generated which are the dimension indices of the feature descriptor (either HOG feature or HOF feature depending on the value of τ.) After that, a "feature difference" can be evaluated with $D_i = x_i^\tau(e_1) - x_i^\tau(e_2), i = 1, 2, \ldots, N_D$. Based on all the D_i, we can estimate the mean and variance of the feature difference.

To put it briefly, a hypothesis (with variables τ, e_1 and e_2) can be generated with the following three steps:

- Generate $\tau \in \{1, 2\}$ to indicate the type of feature to use
- Generate the dimension indexes e_1 and e_2 and compute the feature difference $D_i = x_i^\tau(e_1) - x_i^\tau(e_2), i = 1, 2, \ldots, N_D$
- Split the dataset into two parts based on the mean of feature differences and obtain a variance

We generate γ hypotheses and find the one with the largest variance on feature difference. Since the performance is not sensitive to the number of hypotheses, we fix $\gamma = 50$ for all our experiments. Usually, a larger variance means that the data distribution spreads out more and the feature difference is more significant. Therefore, the corresponding mean is used as the threshold to split the dataset. After this, one node will be built and the dataset will be partitioned into two parts. For each part, a new node will be further constructed in the same way. This process is repeated until the predefined maximum depth is reached.

Compared with the local sensitive hashing (LSH) based indexing used in previous work [26], the benefits of random indexing trees are numerous. In this chapter, we point out four properties that are essential for us. First, each tree in the model is almost independent from others. Second, our random indexing trees are fast to evaluate during the query stage. The computation time only depends on the number of trees and the depth of each tree. Hence, it is usually faster than LSH based nearest neighbor search [26]. In the experiments, we will show that our random-indexing-trees based weighting approach is over 300 times faster than LSH-based approaches. This is of great importance if we want to perform real-time action analysis. Another advantage of random indexing trees compared with LSH is that, during the construction of each tree, data distribution of the STIPs is integrated, which means the tree construction is guided by the data density. This is one reason why random indexing trees has great speed gain but little performance loss. Finally, by adding more trees, we can alleviate the effect of lacking query samples and well model the intra-class variations.

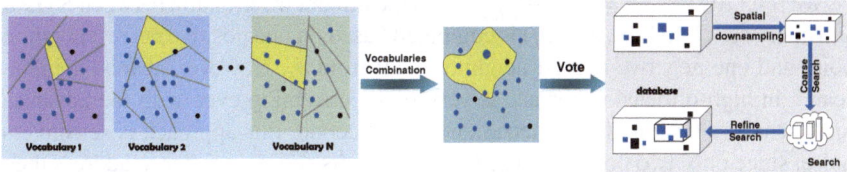

Fig. 3.2 A schematic illustration of randomized-visual-vocabulary indexing and action search

As shown in Fig. 3.2, for each vocabulary, only a small portion of nearest neighbors can be found. The benefit of multiple visual vocabularies is to quickly find enough nearest neighbors to boost the confidences of multiple matches.

3.3 Action Matching Using Randomized Visual Vocabularies

We can consider action search as a template matching process. That is, matching the spatio-temporal template (query STIP points) with all the video sub-volumes in the database. More specifically, our objective is, given one or more query videos, referred to as $\mathcal{Q}$, to extract all the subvolumes which are similar to the query. Formally, that is to find:

$$V^* = \max_{V \subset \mathcal{D}} s(\mathcal{Q}, V), \tag{3.1}$$

where $s(\mathcal{Q}, V)$ is a similarity function between a set of query video clips $\mathcal{Q}$ and a subvolume V in the database.

Unlike previous single template action detection and retrieval [4], which can only take one positive sample for query, our approach can integrate multiple query samples and even negative ones. By introducing negative samples during the query phase, our algorithm is more discriminative. In addition, this approach enables interactive search by leveraging the labels obtained from user feedbacks.

Following our previous work in [26], we use the mutual information as the similarity function for $s(\mathcal{Q}, V)$. So we have:

$$\begin{aligned}
V^* &= \max_{V \subset \mathcal{D}} MI(\mathbf{C} = c_{\mathcal{Q}}, V) \\
&= \max_{V \subset \mathcal{D}} \log \frac{P(V|\mathbf{C}=c_{\mathcal{Q}})}{P(V)} \\
&= \max_{V \subset \mathcal{D}} \log \frac{\prod_{d_i \in V} P(d_i|\mathbf{C}=c_{\mathcal{Q}})}{\prod_{d_i \in V} P(d_i)} \\
&= \max_{V \subset \mathcal{D}} \sum_{d_i \in V} \log \frac{P(d_i|\mathbf{C}=c_{\mathcal{Q}})}{P(d_i)},
\end{aligned} \tag{3.2}$$

where $\mathbf{C} = c_{\mathcal{Q}}$ refers that we want to find the sub-volumes similar to the query clip $c_{\mathcal{Q}}$.

We refer to $s^{c_{\mathcal{Q}}}(d_i) = \log \frac{P(d_i|\mathbf{C}=c_{\mathcal{Q}})}{P(d_i)}$ as the mutual information between STIP d_i and query set $\mathcal{Q}$. In [26], $s^{c_{\mathcal{Q}}}(d_i)$ is computed based on one positive nearest neighbor point and one negative nearest neighbor point from d_i. However, nearest neighbor search in high-dimensional space is very time-consuming even with the advanced local sensitive hashing (LSH) technique [26]. Second, this approach is sensitive to noise, since only two points are used to compute its score. In order to address these problems, we formulate $s^{c_{\mathcal{Q}}}(d_i)$ as:

$$s^{c_Q}(d_i) = \log \frac{P(d_i | \mathbf{C} = c_Q)}{P(d_i)}$$
$$= \log \frac{P(d_i | \mathbf{C} = c_Q) P(\mathbf{C} = c_Q)}{P(d_i) P(\mathbf{C} = c_Q)} \qquad (3.3)$$
$$= \log \frac{P(\mathbf{C} = c_Q | d_i)}{P(\mathbf{C} = c_Q)}.$$

In Eq. 3.3, $P(\mathbf{C} = c_Q)$ is the prior probability that can be computed as the ratio of the number of positive query STIPs to the total number of query STIPs.

In order to estimate $P(\mathbf{C} = c_Q | d_i)$ efficiently and robustly, random indexing trees are used. We consider each tree as one partition of the data space. In the following section, we discuss how to estimate $P(\mathbf{C} = c_Q | d_i)$ given multiple random indexing trees.

Suppose N_T random indexing trees have been built offline from the database. At the query stage, all the STIP points in the query set $Q = Q_P \cup Q_N$ (where Q_P and Q_N refer to positive query and negative query, respectively) are first extracted and distributed into the trees. Figure 3.2 gives a two-dimension example where blue and black dot points represent the positive and negative STIPs, respectively. Each STIP point $d_i \in \mathcal{D}$ (red square in Fig. 3.2) falls into one of the leaves of a tree. Each leaf node contains several STIP points $d_q \in Q$. In order to compute the posterior $P(\mathbf{C} = c_Q | d_i)$, we integrate the information from all the leaves which contain d_i. Suppose d_i falls into a leaf with N_k^+ positive query STIP points and N_k^- negative points for tree T_k, then $P(\mathbf{C} = c_Q | d_i)$ can be computed as:

$$P(\mathbf{C} = c_Q | d_i) = \frac{1}{N_T} \sum_{k=1}^{N_T} \frac{N_k^+}{N_k^+ + N_k^-}. \qquad (3.4)$$

As can be seen from Eq. 3.4, our voting strategy can integrate negative query samples, which makes our algorithm more discriminative.

Equation 3.3 can hence be rewritten as:

$$s^{c_Q}(d_i) = \log P(\mathbf{C} = c_Q | d_i) - \log P(\mathbf{C} = c_Q)$$
$$= \log \frac{1}{N_T} \sum_{k=1}^{N_T} \frac{N_k^+}{N_k^+ + N_k^-} - \log P(\mathbf{C} = c_Q). \qquad (3.5)$$

However, in the case where there are no negative query samples available ($Q_N = \emptyset$), we slightly modify Eq. 3.4 to:

$$P(\mathbf{C} = c_Q | d_i) = \frac{1}{N_T} \sum_{k=1}^{N_T} \frac{N_k^+}{M}, \qquad (3.6)$$

where M is a normalization parameter. And Eq. 3.5 can be written as:

$$
\begin{aligned}
s^{c_Q}(d_i) &= \log \frac{1}{N_T} \sum_{k=1}^{N_T} \frac{N_k^+}{M} - \log P(\mathbf{C} = c_Q) \\
&= \log \frac{1}{N_T} \sum_{k=1}^{N_T} N_k^+ - \log M - \log P(\mathbf{C} = c_Q).
\end{aligned}
\tag{3.7}
$$

We denote $A = -\log M - \log P(\mathbf{C} = c_Q)$. A is a parameter which is set empirically.

Each tree is a partition of the feature space as shown in Fig. 3.2. Hopefully, STIP points in the same leaf node are similar to each other.

The score evaluation (Eq. 3.5) on the trees can be explained intuitively by a dyeing process. We can think of each positive query STIP point as having a blue color and a negative point as having a black color. For each query point, we pass it down each tree. The leaf that the point falls in is dyed in the same color as the query point. Each leaf keeps a count of the number of times it is dyed by blue and a count of the number of times it is dyed by black after we pass all the positive and negative query points down the trees. If a leaf's blue count is larger than the black count, it is more likely to belong to the positive region, and vice versa. Similarly, if the blue count is the same as the black count, it is more likely not to vote any side, i.e., vote zero. Given a point d_i (red square point in Fig. 3.2) in the dataset, to compute its score with respect to the positive queries, we pass it down each tree. From each tree, we find the leaf that d_i falls in. The blue counts and black counts of all the leafs in all the trees that d_i falls in are combined to estimate its posterior $P(\mathbf{C} = c_Q|d_i)$, as given in Eq. 3.5. In some sense, the random indexing trees are like a special kernel, as shown by the yellow regions in Fig. 3.2. The idea of dyeing process is not only limited to trees but also applicable to any vocabulary structure.

Even though our random indexing trees share some similar properties with random forest, e.g. [6, 7], there are important differences between our technique and random forests. First, our random indexing trees are constructed in an unsupervised manner for class-independent video database indexing, while traditional random forests are constructed in a supervised manner. Second, our random indexing trees generate both positive and negative voting scores, thus it is more discriminative compared to [6, 7], which generates only positive votes based on the frequency. Third, we use random indexing trees for density estimation, which has not been exploited before.

3.4 Efficient Action Search

3.4.1 Coarse-to Hierarchical Subvolume Search Scheme

After computing the scores for all the STIP points in the database, we follow the 3D branch and bound approach in [26] to search for subvolumes in each video in the database. The idea of branch and bound search is to branch the search space and give an upper bound to each candidate subset. The candidate subsets can be dropped if the upper bound is lower than the current optimal value. In [26], 3D branch and bound search was proposed and to decompose the search space into the spatial temporal domain, respectively. However, as stated in Chap. 2, there are two limitations in the subvolume search method proposed by [26]. First, we need to run multiple rounds of branch and bound search if we want to detect more tha one instance. In addition, the computational cost is extremely high when the video resolution is high.

In Chap. 2, three speeding up techniques have been proposed to reduce the computational cost of branch-and-bound search. They are spatial-downsampling, λ search and Top-K search. Although significant computational advantages have been achieved, it still takes 26 min to search one hour video database. This is unacceptable for action retrieval application.

To further reduce the computational cost, a coarse-to-fine hierarchical search is proposed here. The basic idea is to first search the coarse resolution score volumes to quickly find a number of candidate subvolumes, and then refine the candidate subvolumes using the finer resolution score volumes. Note that this technique is different from Chap. 2 in that we introduce a refinement mechanism so that the results obtained at a coarser resolution are refined in a higher resolution score volume.

According to Chap. 2, the computational complexity of the top-K search is $O(m^2n^2t) + O(Kmnt)$, where m, n, t are the width, height and duration of the database video and K refers to the top K results. Obviously, the most effective way to reduce the computational cost is to reduce the spatial resolution of the score volume. Thus, spatial-downsampling is performed to compress the search space. The following error bound for score volume downsampling is proposed in Chap. 2.

$$f^s(\tilde{V}^*) \geq (1 - \frac{s * h + s * w + s^2}{wh}) f(V^*), \tag{3.8}$$

where $\tilde{V}^* = \text{argmax}_{V \in \mathcal{D}^s} f^s(V)$ denotes the optimal subvolume in the downsampled search space $\mathcal{D}^s$, $f(V) = MI(Q, V)$, and V^* refers to the optimal subvolume in the original search space with width w and height h. With the help of this error bound, we can relax the top searched list to include more results for a further round of re-ranking. The benefit of using this error bound is to reduce the precision loss in the coarse round of search. Suppose we want to eventually retrieve the top K results from the database, we first downsample the score volumes with a factor of $2a$ and retrieve the top $\hat{K}$ subvolumes by using the top-K search algorithm [25]. We choose $\hat{K} = 2K$ in our experiments.

After that, we can estimate a threshold, denoted as θ, based on the Kth largest subvolume score (denoted as $f^s(\tilde{V}^{(K)})$). For example, if we set downsampling factor $s = 8$ and assume $w = h = 64$, then our approximation has an average error:

$$\frac{s * h + s * w + s^2}{wh} = 56.3 \%. \tag{3.9}$$

So we choose $\theta = 0.437 f^s(\tilde{V}^{(K)})$ to filter the first round results. Then, for each remaining subvolume $\tilde{V}^{(k)}$ from the first round, we first extend the spatial size with 30 pixels in each direction and another round of branch-and-bound search is performed. Different from previous round, which is running over the entire video space, this round of search is performed over the filtered 3D-subvolumes (extended with 30 pixels in each spatial direction). λ search [25] with $\lambda = f^s(\tilde{V}^{(K)})$ and downsampling factor a is used in this round of search.

Despite that only two rounds of search are used, our algorithm can be extended to more rounds of search. This is especially useful to handle high-resolution videos. As given in Table 3.5, our efficient two-round branch-and-bound search only costs 24.1 s to search a database of 1 h long 320×240 videos. This is over 60 times faster than the approach in Chap. 2 and 2,900 times faster than [26]. Even for a 5 h large database, it only costs 37.6 seconds to respond to the users. Besides the speed advantages, we will see from the experiments that the search accuracy is not compromised.

Finally, we differentiate our work from Chap. 2 here. In Chap. 2, only one round of search is used. The search is performed in the entire three-dimensional video space. However, in our work, with the help of the coarse round search, our fine round search only need to focus on the potential regions, which can save significant computational cost. Besides, to reduce the computational loss from the coarse round search, the error bound is used to determine the number of candidate subvolumes for re-ranking.

3.4.2 Refinement with Hough Voting

Although our search algorithm can successfully locate the retrieved actions, the localization step may not be accurate enough, as can be seen from the first row of Fig. 3.4. This motivates us to add a refinement step. The idea is to back-project the initial subvolume into the query video. Based on the matches of STIPs, we can vote the action center (only in the spatial domain). Figure 3.3 is an illustration of the Hough refinement step.

Suppose we already have the initial results from the down-sampled branch and bound search (the blue region in the left image of Fig. 3.3), for all the STIP points within the detected subvolume, we match them with the STIPs in the query video clip, either by trees or Nearest Neighbor search. Then the shift from the matched STIPs in the query will vote for the center of the retrieved action. To simplify the problem, we only consider one fixed scale and smooth the votes with a Gaussian kernel. After considering all the votes, the center of the retrieved action is the position with the largest vote (the red cross in the third image of Fig. 3.3). To recover the spatial extent, we set the spatial scale of the action to be the smallest subvolume which includes the initial retrieved region and the temporal scale is fixed to the initial retrieved result.

After the refinement, the blue region in the right image of Fig. 3.3 gives an illustration of the revised result compared with original round of result, i.e., the left image. Both quantitative results, as shown in Fig. 3.7, and empirical results in Fig. 3.4 show

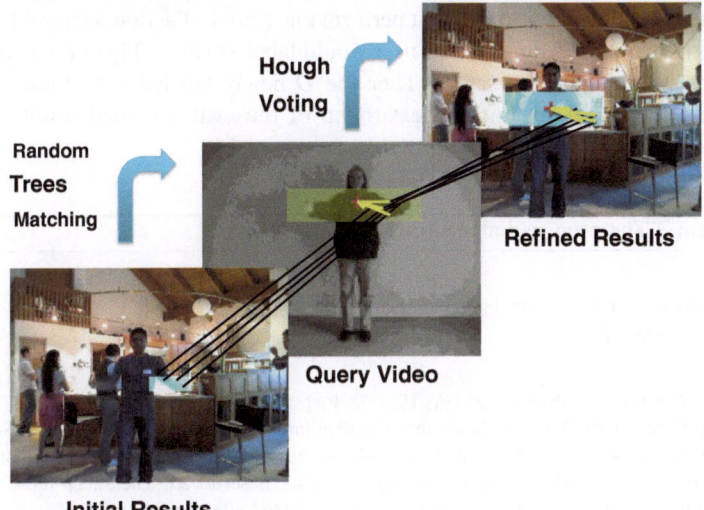

Fig. 3.3 Illustration for Hough refinement

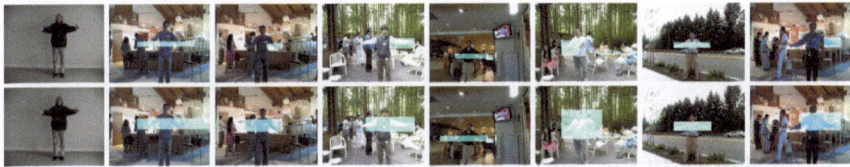

Fig. 3.4 Comparison of retrieval results without and with Hough refinement. For each row, the first image indicates the query sample and the following seven images refer to the highest ranked retrieved results. All the experiments are done with only one query sample without user feedback. The *upper* and *lower rows* are the experiments on handclapping without and with Hough refinement, respectively. The query clip (*first column*) is from KTH while the database is from MSR II

that the refinement step can successfully improve our retrieved results. The major steps of our algorithm are described at Algorithm 3.

3.4.3 Interactive Search

The performance of our action retrieval system is constrained by the limited number of queries. To show that our retrieval system can achieve better results when more queries are provided, we add an interaction step to facilitate human interaction. There are two major advantages of the interaction step. The first is to allow the user to express what kind of action he/she wants to retrieve. Another advantage is that our system can benefit from more query samples after each round of interaction.

To implement the system, we first perform one round of action retrieval based on a few query samples. After that, the user would label D ($D = 3$ in our experiments) detections with the highest scores. Then the D newly labeled subvolumes will be added into the query set for the next round of retrieval. Detailed results will be discussed in the experiment section.

Algorithm 3 The proposed algorithm for action search

Require:

- Database with the random indexing trees: $\{T_k\}$
- Query video: Q

Ensure:
 The top K retrieved sub-volumes $\{V_1, V_2, \cdots, V_K\}$
1: **Voting:** Based on Eq. 3.7, vote the database STIP points by the random indexing trees (Sect. 3.3).
2: **Coarse-round search:** Retrieve the top $\hat{K}$ subvolumes by using the top-K search algorithm for each video in the database. The downsampling factor is set to $2a$. Then keep those detected sub-volumes with scores above the threshold θ (discussed after Eq. 3.9) only.
3: **Fine-round search:** Rerank the remaining subvolumes from the first round of results with branch and bound search (downsampling factor a).
4: **Refinement:** [optional] Hough refinement discussed in Sect. 3.4.2.

3.4.4 Computational Complexity

For our action retrieval system, there are two major runtime costs: voting and searching. The computation complexity is $O(N_s T_d N_T)$ for the voting step where N_s refers to the number STIPs in a query clip, T_d refers to tree depth, and N_T refers to the number trees in a forest. As shown in the Table 3.5, the voting time is negligible compared to the search time. For action search, the worst time complexity is

$$\begin{aligned}
T = & O((m/2a)^2(n/2a)^2 t) + O(\hat{K}(m/2a)(n/2a)t) \\
& + O((\hat{m}/a)^2(\hat{n}/a)^2\hat{t}) + O(K(\hat{m}/a)(\hat{n}/a)\hat{t})
\end{aligned} \tag{3.10}$$

where m, n and t are width, height and duration of the clips in database. a is the downsampling factor and $\hat{K}$ (this value depends on the retrieval scores) is a little larger than K ($K = 7$ refers to the number of retrieved results in our experiment). After a filtering step (the first two complexity), $\hat{m}, \hat{n}, \hat{t}$ are used to represent the spatial width, spatial height and temporal duration for remaining sequences (usually $\hat{t} \ll t$), respectively. The quantitative analysis of the computational cost will be discussed in the experimental part.

3.5 Experimental Results

To validate our proposed algorithm, seven experiments on six datasets have been discussed in this section. KTH [12] and MSR II [26] are used to evaluate our algorithm for action classification and action detection, respectively. Table 3.3 lists the five datasets for the validation of our search algorithms. Sample frames from the five datasets can be found in Fig. 3.5. Since our algorithm is trying to handle the action search problem, we focus on the action search experiments in this section. In order to provide a quantitative comparison with other work, we give an action retrieval experiment on benchmark dataset MSRII first. After that, illustrative experiments for action search on CMU dataset [9], Youtube videos, and UCF dataset [19] are discussed. To show the ability to handle real action retrieval by our system, we build a 5 h dataset with videos downloaded from datasets MSRII [26], CMU [9], VIRAT [17], Hollywood [14] and some videos downloaded from Youtube.

Fig. 3.5 Sample frames from our testing datasets. The first to the third rows show sample frames from MSR II, CMU, and Youtube videos, respectively. The fourth row shows some different frames from the 5 h large dataset

Table 3.1 Confusion matrix for KTH action dataset. The total accuracy is 90.4 %

	Walk	Clap	Wave	Box	Run	Jog
Walk	142	0	0	0	0	2
Clap	0	136	1	7	0	0
Wave	0	11	133	0	0	0
Box	4	0	0	140	0	0
Run	1	0	0	0	114	29
Jog	2	0	0	0	26	116

Table 3.2 Comparison of different reported results on KTH dataset

Method	Average accuracy (%)
Our method	90.4
[25]	91.8
[18]	90.3
[12]	91.8
[3]	95.02
[15]	94.4
[11]	94.53

Table 3.3 List of datasets for experiments

Dataset	Total length	Query action
MSR II	1 h	Waving, clapping, boxing
CMU	20 min	Waving, bending
Youtube	4.5 min	tennis serve
UCF sports	≈15 min	Diving, weightlifting
Large dataset	5 h	Waving, clapping, boxing, ballet spin

3.5.1 Action Classification on KTH

We first give an experiment to show that our algorithm is able to handle the traditional action classification problem. We use the benchmarked KTH dataset [12] and test our algorithm with the same setting as in [12, 25]: $8+8$ sequences for training/validation and nine sequences for testing. Table 3.1 lists the confusion matrix for our algorithm and a comparison with other works are listed in Table 3.2. Although our work aims to handle the action search problem, i.e., the label information from the training data is not utilized until the testing stage, we still achieve comparable results according to Table 3.2.

3.5.2 Action Detection on MSR II

We then validate our random-indexing-trees strategy with a challenging action detection experiment. Since handwaving is an action that is quite common in real life, we choose to detect handwaving actions in this experiment. We first train the model with KTH dataset (with 16 persons in the training part) and then perform experiments on a challenging dataset (MSR II) of 54 video sequences, where each video consists of several actions performed by different people in a crowded environment. Each video is approximately 1 min long.

Figure 3.6 compares the precision-recall curves for the following methods (the resolution for the original videos is 320×240):

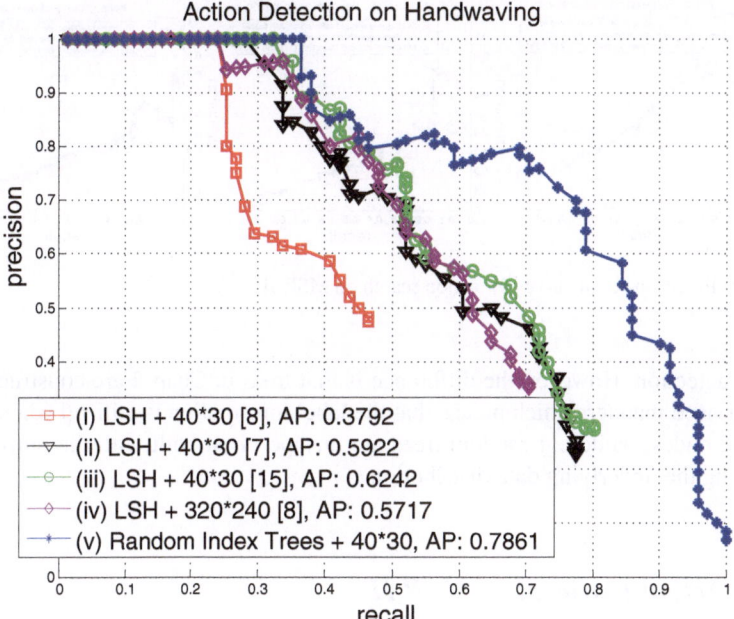

Fig. 3.6 Precision-recall curves for handwaving detection on MSR II. AP in the legend means the average precision

(i) Accelerated Spatio-Temporal Branch-and-Bound search (ASTBB) [27] in low resolution score volume (frame size 40×30),

(ii) Multi-round branch-and-bound search [26] in low-resolution score volume (frame size 40×30),

(iii) Top-K search in down-sampled score volume discussed in Chap. 2 (size 40×30, but for the indexing we do not use the supervised trees in Chap. 2,

(iv) ASTBB [27] in 320×240 videos,

(v) Random-indexing-trees based voting followed by Top-K search in down-sampled score volume (size 40×30).

The first four methods ((i)–(iv)) use the LSH based voting strategy [26]. The measurement of precision and recall is the same as those described in [26]. To compute the precision we consider a true detection if : $\frac{\text{Volume}(V^* \cap G)}{\text{Volume}(G)} > \frac{1}{8}$, where G is the annotated ground truth subvolume, and V^* is the detected subvolume. On the other hand, to compute the recall we consider a hit if: $\frac{\text{Volume}(V^* \cap G)}{\text{Volume}(V^*)} > \frac{1}{8}$. According to Fig. 3.6, our random-indexing-trees based action detection outperforms the other algorithms. Compared with LSH voting strategy ((i)–(iv)), it shows that our random-indexing-trees based voting is more discriminative and robust. The underlying reason is that our random trees are data-aware, i.e., we model the data distribution when constructing the trees. Besides, since LSH only uses two nearest neighbors for voting, the results are easily corrupted by noise. In Chap. 2, random forest is used to perform

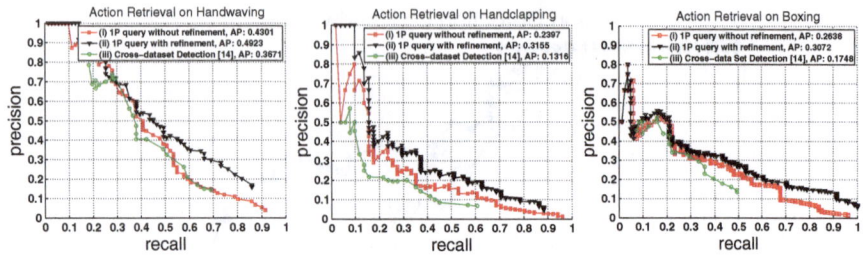

Fig. 3.7 Precision-recall curves for action search on MSR II

action detection. However, the difference is that trees in Chap. 2 are constructed in a supervised manner, which means that the label information is utilized when splitting the nodes, while our random trees are unsupervised built with the purpose of modeling the underlying data distribution.

3.5.3 Action Retrieval on MSR II

To give a quantitative result for our action retrieval system, we use videos from MSR II as the database. The query samples are drawn from KTH dataset. As there have not been any reported action retrieval results on MSR II dataset, we compare our retrieval results with several previously reported action detection results on this dataset. The evaluation is the same as that for action detection. For the implementations of our random indexing trees, we set the number of trees in a forest $N_T = 550$ and the maximum tree depth to 18. Figure 3.7 compares the following three strategies on handwaving, handclapping and boxing actions (for the boxing action, we flip each frame in the query video so that we can retrieve the boxing coming from both directions[1]), respectively.

 (i)   One positive query example without Hough refinement,
 (ii) One positive query example with Hough refinement,
 (iii) Cross-Dataset detection [3],[2]

As shown in Fig. 3.7, with a single query, our results ((i) and (ii)) are already comparable to (iii) for all three action types. This is quite encouraging because (iii) used all the training data while we only use a single query. Besides, our Hough refinement scheme (ii) improves the results without Hough refinement (i).

---

[1] Only for the boxing action, we use the query video as well as the flipped version. For other actions, only the query video is used.

[2] The STIP features in [3] are extracted in video resolution of 160×120 but 320×240 for other methods.

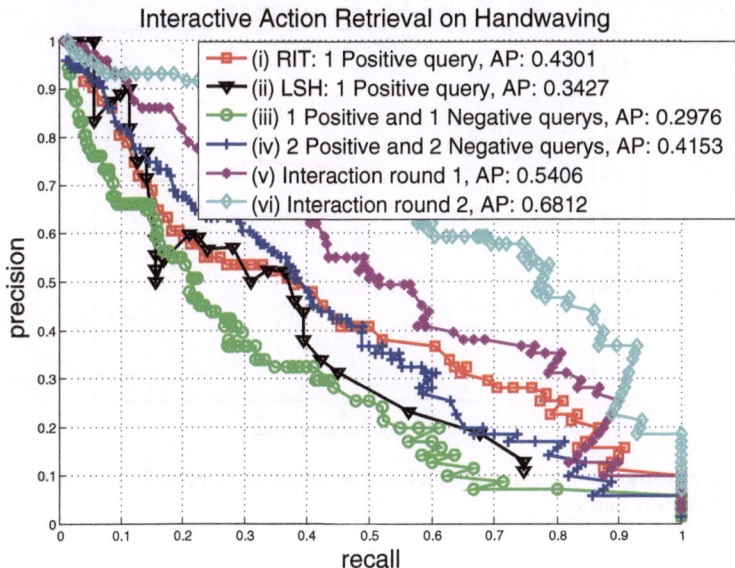

Fig. 3.8 Precision-recall curves for the interactive action retrieval on MSR II

Figure 3.8 shows the experimental results of interactive action retrieval. The following six strategies (all of them are performed without Hough refinement) are compared.

(i)   One query example with random indexing trees (RIT) based voting,
(ii) One query example with LSH based voting,
(iii) One positive and one negative query examples
(iv) Two positive and two negative query examples,
(v)   One iteration of user interaction after (i),
(vi) Two iterations of user interaction after (i).

Performance for LSH-based indexing scheme [26] is listed with the similar framework as random indexing trees. The parameters for LSH are set to make the comparison fair. We can see that when there is only one query example, our random-indexing-trees based voting strategy (i) is superior to LSH based voting strategy (ii). When there are two query examples (one positive and one negative,) the retrieval results become worse than the one query case. The reason is that negative action type is more difficult to describe and a single negative example is sometimes not enough. Figure 3.9 shows that we can increase our average precision by increasing the number of negative queries. The results would be stable if we use around four negative queries along with one positive query. Besides, the performance of our system increases as more query samples are given by interaction. In particular, after two interaction steps, our retrieval results are better than the results obtained by other

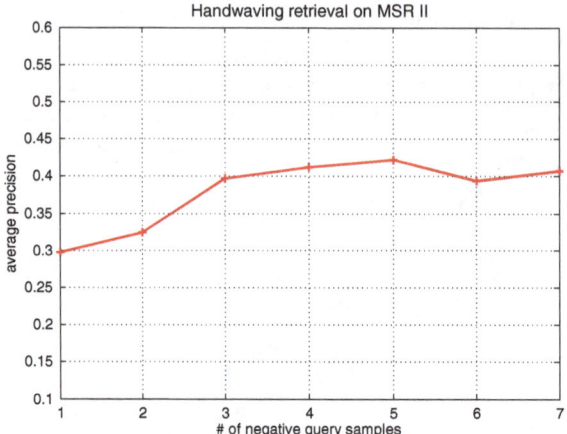

Fig. 3.9 Comparison of average precision based on the input of one positive query but different number of negative queries

action detection systems ((i)–(iv) in Fig. 3.6), which utilize all the training data (256 examples).

We also provide some illustrative results in Fig. 3.4. For each query, seven subvolumes with the highest scores are listed in the figure. The retrieved subvolumes are marked by colored rectangles. The rectangle with cyan background indicates a "correct" retrieval. As shown in the first row of Fig. 3.4, some of the cyan results are focused on a subregion of the action region. But this can be relieved with Hough refinement as indicated in the second row. In short, our action retrieval system can get very good results among the top retrieved subvolumes on various actions types.

3.5.4 Action Retrieval on CMU Database

CMU database [9] is another widely used database for action analysis. Since the annotation of the actions includes the entire human rather than the action itself (as can be seen from Fig. 3.10, our results only mark the region where the action happens), we only give some illustrative examples on this dataset. The CMU database includes 48 videos of total duration around 20 min. The resolution for these videos are 160×120. Handwaving and bending actions are retrieved from the database where the query video for handwaving is from KTH and the query video for bending is from Weizmann dataset [8].

Figure 3.10 shows the search results. For each row, the first image is a sample frame from the query video and the following seven images are from the top-7 retrieved segments, respectively. The cyan region shows the positive detection while the yellow region shows the negative detection. Compared with handwaving, bending is a

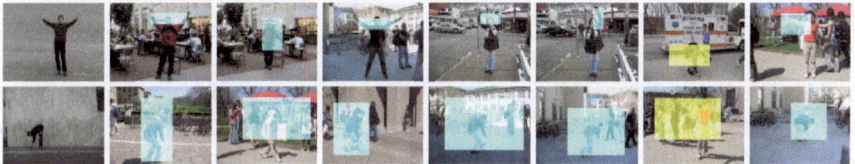

Fig. 3.10 Retrieval results for CMU dataset. For each row, the first image indicates the query sample and the following seven images refer to the highest ranked retrieved results. All the experiments are done with only one query sample without any user feedback. For the *first row*, the query clip with handwaving action is from KTH while the database is from CMU [9]. For the *second row*, the query clip with bending action is from Weizmann Dataset while the database is from CMU [9]

Fig. 3.11 An illustrative example to search tennis serve action. The query and database videos are downloaded from Youtube. The seven images in the *first row* show different frames in the query video while the images from 2rd to 6th rows show the top-5 searched results

non-periodic action, which is more challenging due to simple motion pattern and small number of query STIP points. From this experiment, we can see that our algorithm can handle the two actions with a large intra-class variations and clutter background, even in the low-resolution and highly compressed videos.

3.5.5 Action Retrieval on Youtube Video

In this experiment, we validate our algorithm with a challenging tennis serve action search from a Youtube video,[3] which is also a nonperiodic action. More action searches from Youtube videos will be available from our project website. The length for the database video is around 280 s, with several tennis serving actions performed by different actors under different views. The query video is a 2 s segment cut from a different Youtube video.[4] The experiment is very challenging due to the following aspects. First, there are different scenes and players compared with the query clip. Besides, the serving actions are recorded in several different views. Second, it contains not only the serving action but also a lot of other actions as well. We need to differentiate the serving action from other actions. Third, in addition to the large intra-class variations, the visual quality is poor due to video compression. The regions marked blue (the reason to use blue color is that it differentiates with the background color) from 2rd to 6th rows of Fig. 3.11 are the top 5 retrieved subvolumes based on the query video from the first row. We can see that our algorithm achieves promising results.

3.5.6 Action Retrieval on UCF Sports Database

We further validate our algorithm with UCF sports dataset [19]. We choose to search diving and weightlifting actions because there are no large camera motions. Since the videos have already been segmented, no localization is needed. Figure 3.12 shows our experimental result. The two rows illustrate diving and weightlifting actions, respectively. For each row, the first column is a sample frame from the query video, and the subsequent five columns are the top-5 retrieved results. Although only a single query sample is used for search, we still obtain quite promising results.

Fig. 3.12 Search results of diving and weightlifting on UCF Sports dataset. The *first column* is from the query video and the following six columns refer to the top-5 retrieved results. The fourth and fifth results for the diving action and the second result for weightlifting are false positives caused by similar motion patterns

---

[3] http://www.youtube.com/watch?v=inRRaudOf5g.

[4] http://www.youtube.com/watch?v=NQcmYTIrqNI.

3.5.7 Action Retrieval on Large-Scale Database

To verify that our algorithms can handle large scale dataset, we build a large database with more than 200 videos. The database includes videos from datasets MSRII [26], CMU [9], VIRAT [17], Hollywood [14] and some videos downloaded from Youtube. The total duration is around 5 hours.

Four different actions (handwaving, handclapping, boxing and ballet spinning) are tested in this large dataset. Each experiment is done with only one query video, without any post-processing, e.g. Hough refinement. The query videos for the first three actions (around 15 s for each action) are collected from KTH while the query video for ballet spinning is downloaded from Youtube (around 5 s). For handwaving, handclapping, and boxing, we retrieve top-40 detections. For ballet spin, we retrieve top-10 detections since there are not as many ballet spin actions in the database. Figure 3.13 shows five samples of the retrieved results of handwaving. The first row gives seven frames from the query video while the second to fifth rows show the four positive results where the retrieved subvolumes are marked with cyan color. The sixth row shows one negative result where the retrieved subvolumes are marked with yellow color. Similarly, Figs. 3.14, 3.15 and 3.16 show the results of handclapping, boxing and ballet spin, respectively. Besides, in Fig. 3.17, we give the retrieved performance (precision versus the number of top samples retrieved) for the large database. Based

Fig. 3.13 Illustrations for handwaving retrieval in large dataset. First row shows the seven frames from query video while the following rows give the five retrieved examples (four positive examples marked by *cyan color* and one negative example marked by *yellow* color)

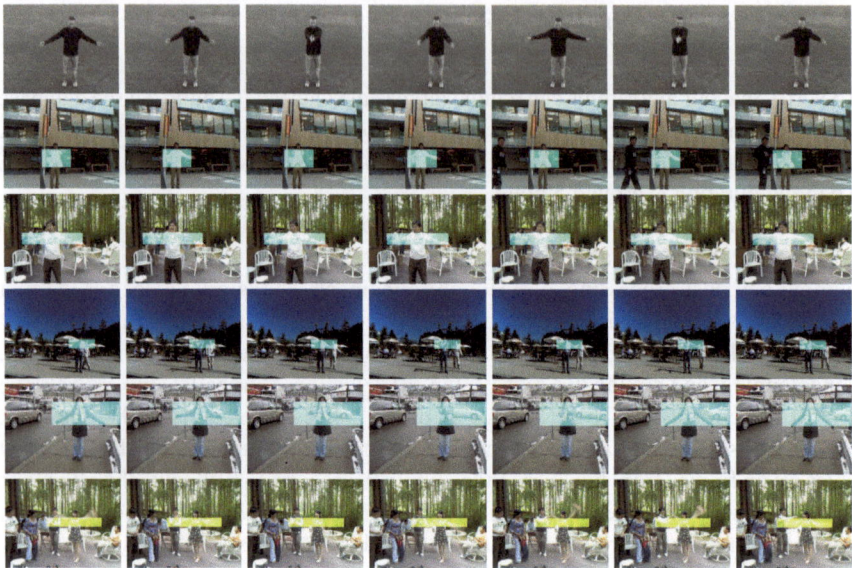

Fig. 3.14 Illustrations for handclapping retrieval in large dataset. First row shows the seven frames from query video while the following rows give the five retrieved examples (four positive examples marked by *cyan color* and one negative example marked by *yellow color*)

on the illustrative results, we can see that our algorithm can well handle the large scale changes, clutter background, partial occlusion and low visual quality.

3.5.8 Implementation Issues

To implement such a system, there are several issues we need to take care of in both the indexing stage and query stage. For indexing part, we need to determine the number of trees. We use an experiment to evaluate the relationship between the number of trees and average precision. The test environment is the same as that discussed in Sect. 3.5.3. According to Fig. 3.18, the number of trees become stable from 300 for the handwaving and boxing action and from 500 for the handclapping action. Besides, if we use only one tree (traditional BOW model), the system cannot find any positive detections. The reason is that only a very limited number of database STIP points are matched to the query STIP points and these matched STIP points are isolated from each other. Usually only the STIP point itself forms a detection. Hence, we need to increase the number of trees to introduce more matches. But as shown in the figure, the performance will be stable when sufficient number of trees are used.

Fig. 3.15 Illustrations for boxing retrieval in large dataset. *First row* shows the seven frames from query video while the following rows give the five retrieved examples (four positive examples marked by *cyan color* and one negative example marked by *yellow color*)

Based on Fig. 3.18, we fix the number of vocabularies (trees) as 550 in our experiment setting. The depth for the trees is set as 18 for most of our experiments (For Youtube video dataset, we set it as 15 because the database is of small size). In the query stage, the downsampling factor of the branch-and-bound search (referring as a) is first set to 16 in the coarse round of search and then refined as 8 for another round of search.

3.5.9 Computational Cost

In offline stage, we need to first extract the STIP features from the database videos. The time cost depends on the content of the videos, video resolution and video duration. The code for STIP feature is downloaded from the author's website [13]. After that, it costs around 3 hours to build a vocabulary with 550 trees on the STIP features for MSR II dataset. We do not perform other pre-processing on the dataset. For the online cost, there are two major runtime costs: voting and searching. The total computational cost for our system is listed in Table 3.5. The testing environment is as follows. We use one query video, which is approximately 20 s long. Two datasets are tested in our experiments. The first database is MSR II, which consists of 54 sequences with 320×240 resolution. The second dataset contains 5 h of videos

Fig. 3.16 Illustrations for ballet spin retrieval in large dataset. *First row* shows the seven frames from query video while the following rows give the five retrieved examples (four positive examples marked by *cyan color* and one negative example marked by *yellow color*)

(discussed in Sect. 3.5.7). To set the parameter θ in Sect. 3.4.1, we average the w and h among the top K results and obtain an error bound based on the estimated w and h. With this error bound, we can compute θ similarly as in Eq. 3.9. We use a PC with 3GHz CPU and 3G memory. Table 3.4 compares the voting cost: the random-indexing-trees based vocabulary implementation is much more efficient compared with LSH. According to Table 3.4, random-indexing-trees method is over 300 times faster than LSH based indexing but with even superior performance from Fig. 3.8. For the searching cost, as shown in Table 3.5, our coarse-to-fine subvolume search scheme only costs 24.1s for all 54 video clips in MSR II, while Top-K search in Chap. 2 takes 26min. This is even 2,800 times faster than the branch and bound search in [26]. For the 5h large dataset, it only costs 37.6s to retrieve the top-7 results. From the statistics of Table 3.5, we can see that the increase of database size (from 1 to 5h) do not significantly increase the computational cost (from 26.4 to 37.6s). The reason is that the search consists of two rounds: coarse search and fine search. The fine search time is almost the same for the two datasets since we only consider the similar number of candidates received from the coarse round search. On the other hand, due to the high down-sampling factor in the coarse round search, the computational burden for large dataset is not that intensive.

Note that the total computation time is independent of the duration of the query videos. This means, when there are more queries, the total computation time only grows linearly with the feature extraction time, which is around 30s for a 20s sequence. For a very large database, like Youtube, it has little impact to our

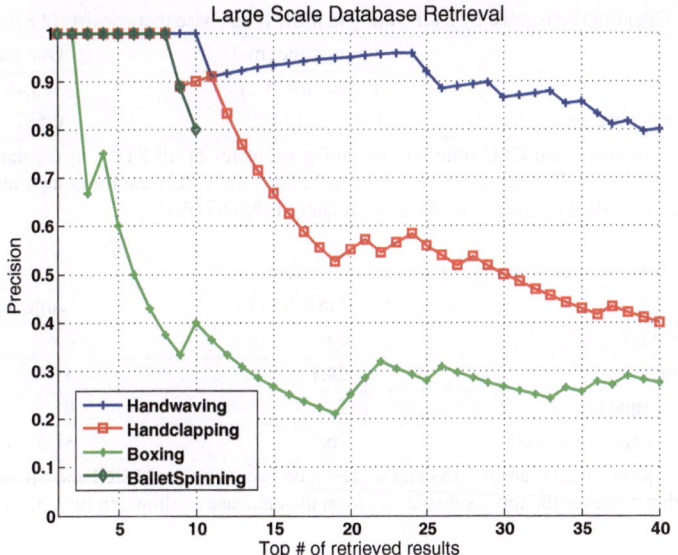

Fig. 3.17 Retrieval results from large scale dataset

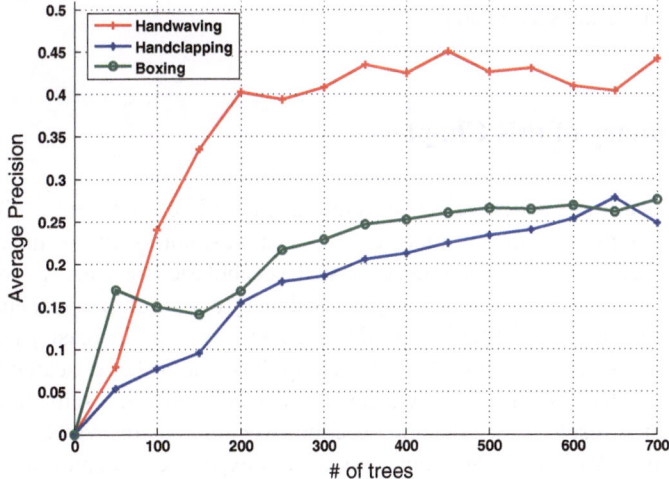

Fig. 3.18 Relation between number of trees and average precision

voting cost since the voting cost mainly depends on the number of trees and the depth of each tree. In order to deal with the increasing search complexity, parallel computing can be utilized in the first step of branch and bound search since the search for different video clips are mutually independent. As the number of candidates for

Table 3.4 CPU time consumed by STIP voting in MSR II database that consists of 870,000 STIPs

Method	Voting time (ms)	One sequence (s)
LSH [26]	173.48 ± 423.71	173.48
Random indexing trees	0.537 ± 0.14	0.537

The second column is the CPU time for computing the votes of all STIPs in the database with respect to a single STIP in the query. The third column is the CPU time for computing the votes with respect to a 10 s long query video (approximately 1,000 STIPs)

Table 3.5 Total computation time of our retrieval system

Dataset	MSR II (1 h)	Large dataset (5 h)
Voting time (s)	0.6	0.6
Search time (s)	24.1	37
Refinement time (s)	2	0
Total Computation Time (s)	26.7	37.6

Suppose the query video is around 15 s and we test it on two database: MSRII and 5 h large dataset. Our algorithm retrieves the top 7 subvolumes from the database as shown in Figs. 3.4 and 3.10

search in the second step of our branch and bound search only depends on the number of retrieved results required by the user, the database size has little impact on the runtime cost for the second step.

3.6 Summary of this Chapter

We developed a fast action search system that can efficiently locate similar action instances to a query action. To index the video interest points for fast matching, we proposed to build multiple randomized-visual-vocabularies by using random indexing trees. Compared with using a single vocabulary tree, multiple vocabulary trees better compensate for information loss due to quantization. By increasing the number of vocabularies, we can improve the matching thus lead to better search accuracy. To achieve faster response time, we developed a coarse-to-fine subvolume search scheme which results in a dramatic speedup over the existing video branch-and-bound method. Various challenging cross-dataset experiments demonstrate that our proposed method is not only fast to search large-scale video dataset, but also robust to action variations, partial occlusions, and cluttered and dynamic backgrounds. Moreover, our technique has the unique property that it is easy to leverage feedbacks from the user.

References

1. E. Boyer, D. Weinland, R. Ronfard, Free viewpoint action recognition using motion history volumes. Comput. Vis. Image Underst. **104**(2–3), 207–229 (2006)
2. L. Breiman, Random forests. Mach. Learn. **45**, 5–32 (2001)
3. L. Cao, Z. Liu, T.S. Huang, Cross-dataset action recognition, in *Proceedings of the IEEE Conference on Computer Vision and Pattern Recognition*, pp. 1998–2005 (2010)
4. K.G. Derpanis, M. Sizintsev, K. Cannons, R.P. Wildes, Efficient action spotting based on a spacetime oriented structure representation, in *IEEE Conference on Computer Vision and Pattern Recognition*, pp. 1990–1997 (2010)
5. L. Duan, D. Xu, I.W. Tsang, J. Luo, Visual event recognition in videos by learning from web data, in *IEEE Conference on Computer Vision and Pattern Recognition*, pp. 1959–1966 (2010)
6. J. Gall, V. Lempitsky, Class-specific hough forests for object detection, in *Proceedings of the IEEE Conference on Computer Vision and Pattern Recognition*, pp. 1022–1029 (2009)
7. J. Gall, A. Yao, N. Razavi, L. Van Gool, V. Lempitsky, "Hough forests for object detection, tracking, and action recognition", in *IEEE Transaction on Pattern Analysis and Machine Intelligence*, pp. 2188–2202 (2011)
8. L. Gorelick, M. Blank, E. Shechtman, M. Irani, R. Basri, Actions as space-time shapes. TPAMI **29**, 2247–2253 (2007)
9. Y. Ke, R. Sukthankar, and M. Hebert, "Event detection in crowded videos", in *Proceedings of the IEEE International Conference on Computer Vision*, pp. 1–8 (2007)
10. Y. Ke, R. Sukthankar, M. Hebert, Efficient visual event detection using volumetric features, in *Proceedings of the IEEE International Conference on Computer Vision*, pp. 166–173 (2005)
11. A. Kovashka, K. Grauman, Learning a hierarchy of discriminative space-time neighborhood features for human action recognition, in *IEEE Conference on Computer Vision and Pattern Recognition*, pp. 2046–2053 (2010)
12. I. Laptev, M. Marszalek, C. Schmid, B. Rozenfeld, Learning realistic human actions from movies, in *Proceedings of the IEEE Conference on Computer Vision and Pattern Recognition*, pp. 1–8 (2008)
13. I. Laptev, On space-time interest points. Int. J. Comput. Vis. **64**(2–3), 107–123 (2005)
14. M. Marszałek, I. Laptev, C. Schmid, Actions in context, in *CVPR* (2009)
15. R. Minhas, A.A. Mohammed, Q.M. Jonathan Wu, Incremental learning in human action recognition based on snippets. IEEE Trans. CSVT **22**, 1529–1541 (2012)
16. J. Niebles, H. Wang, L. Fei-Fei, Unsupervised learning of human action categories using spatial-temporal words. Int. J. Comput. Vis. **3**, 299–318 (2008)
17. S. Oh, A. Hoogs, A.G. Amitha Perera, et al., A large-scale benchmark dataset for event recognition in surveillance video, in *CVPR* (2011)
18. K.K. Reddy, J. Liu, M. Shah, Incremental action recognition using feature-tree, in *Proceedings of the IEEE International Conference on Computer Vision*, pp. 1010–1017 (2009)
19. M.D. Rodriguez, J. Ahmed, M. Shah, Action mach a spatio-temporal maximum average correlation height filter for action recognition, in *Proceedings of the IEEE Conference on Computer Vision and Pattern Recognition*, pp. 1–8 (2008)
20. C. Schuldt, I. Laptev, B. Caputo, Recognizing human actions: a local svm approach, in *Proceedings of the IEEE Conference on Pattern Recognition*, pp. 57–62 (2004)
21. Y. Xie, H. Chang, Z. Li, L. Liang, X. Chen, D. Zhao, A unified framework for locating and recognizing human actions, in *CVPR* (2011)
22. G. Yu, J. Yuan, Z. Liu, Propagative hough voting for human activity recognition, *Proceedings of the European Conference on Computer Vision* (*ECCV*) (2012)
23. G. Yu, J. Yuan, Z. Liu, Real-time human action search using random forest based hough voting, in *ACM Multimedia* (2011)
24. G. Yu, J. Yuan, Z. Liu, Unsupervised random forest indexing for fast action search, in *CVPR* (2011)
25. G. Yu, A. Norberto, J. Yuan, Z. Liu, Fast action detection via discriminative random forest voting and top-K subvolume search. IEEE Trans. on Multim. **13**(3), 507–517 (2011)

26. J. Yuan, Z. Liu, Y. Wu, Discriminative subvolume search for efficient action detection, in *Proceedings of the IEEE Conference on Computer Vision and Pattern Recognition*, pp. 2442–2449 (2009)
27. J. Yuan, Z. Liu, Y. Wu, Z. Zhang, Speeding up spatio-temporal sliding-window search for efficient event detection in crowded videos, in *ACM Multimeida Workshop on Events in Multimedia*, pp. 1–8, (2009)

Chapter 4
Propagative Hough Voting to Leverage Contextual Information

Abstract Generalized Hough voting has shown promising results in both object and action detection. However, most existing Hough voting methods will suffer when insufficient training data are provided. To address this limitation, we propose propagative Hough voting in this chapter. Instead of training a discriminative classifier for local feature voting, we first match labeled feature points to unlabeled feature points, then propagate the label and sptatio-temporal configuration information via Hough voting. To enable a fast and robust matching, we index the unlabeled data using random projection trees (RPT). RPT can leverage the low-dimension manifold structure to provide adaptive local feature matching. Moreover, as the RPT index can be built in either labeled or unlabeled dataset, it can be applied to different tasks such as action search (limited training) and recognition (sufficient training). The superior performances on benchmarked datasets validate that our propagative Hough voting can outperform state-of-the-art techniques in various action analysis tasks, such as action search and recognition.

Keywords Action recognition · Propagative hough voting · Random projection tree · Limited training data · Action search

4.1 Introduction

In Chaps. 2 and 3, the computational issue for human action analysis with spatio-temporal localization has been well addressed. However, the recognition and search accuracy have been compromised to reduce the computational cost. In this chapter, we will present a novel algorithm which significantly improves the recognition and search results with only a little computational burden compared with the algorithm in Chap. 3. Instead of dropping the spatio-temporal configuration of local interest points as in Chap. 3, we are trying to leverage this information by Hough voting.

© The Author(s) 2015   57
G. Yu et al., *Human Action Analysis with Randomized Trees*,
SpringerBriefs in Signal Processing, DOI 10.1007/978-981-287-167-1_4

Hough-transform-based local feature voting has shown promising results in both object and activity detections. It leverages the ensemble of local features, where each local feature votes individually to the hypothesis, thus can provide robust detection results even when the target object is partially occluded. Meanwhile, it takes the spatial or spatio-temporal configurations of the local features into consideration, thus can provide reliable detection in the cluttered scenes, and can well handle rotation or scale variation of the target object.

Despite previous successes, most current Hough-voting based detection approaches require sufficient training data to enable discriminative voting of local patches. For example, [4] requires sufficient labeled local features to train the random forest. When limited training examples are provided, e.g., given one or few query examples to search similar activity instances, the performance of previous methods is likely to suffer. The root of this problem lies on the challenges of matching local features to the training model. Due to the possibly large variations of activity patterns, if limited training examples are provided, it is difficult to tell whether a given local feature belongs to the target activity or not. Thus, the inaccurate voting scores will degrade the detection performance.

In this chapter, we propose *propagative Hough voting* for activity analysis. To improve the local feature point matching, we introduce random projection trees [2], which are able to capture the intrinsic low-dimensional manifold structure to improve matching in high-dimensional space. With the improved matching, the voting weight for each matched feature point pair can be computed more reliably. Besides, as the number of trees grows, our propagative Hough voting algorithm is theoretically guaranteed to converge to the optimal detection.

Another nice property of the random projection tree is that its construction is unsupervised, thus making it perfectly suitable for leveraging the unlabeled test data. When the amount of training data is small such as in activity search [19, 22] with one or few queries, one ca use the test data to construct the random projection trees. After the random projection trees are constructed, the label information in the training data can then be propagated to the testing data by the trees.

Our method is explained in Fig. 4.1. For each local patch (or feature) from the training example, it searches for the best matches through RPT. Once the matches in the testing video are found, the label and spatial-temporal configuration information are propagated from the training data to the testing data. The accumulated Hough voting score can be used for recognition [14, 18] and detection [7, 21]. By applying the random projection trees, our proposed method is as efficient as the existing Hough-voting-based activity recognition approach, e.g., the random forest used in [4]. However, our method does not rely on human detection and tracking, and can well handle the intra-class variations of the activity patterns. With an iterative scale refinement procedure, our method can handle small-scale variations of activities as well.

We evaluate our method in two benchmarked datasets, UT-interaction [16] and TV Human Interaction [12]. To fairly compare with existing methods, we test our propagative Hough voting with (1) insufficient training data, e.g., in activity search

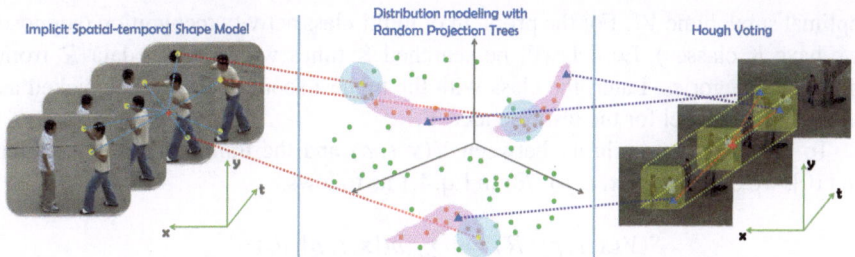

Fig. 4.1 *Propagative Hough Voting* The *left* figure illustrates our implicit spatial-temporal shape model on a training video. Three sample STIPs from the testing videos are illustrated with *blue triangles* in the *right* figure. Several matches will be found for the three STIPs given the RPT. We choose three *yellow dots* to describe the matched STIPs from the training data in the *middle* figure. For each training STIP (*yellow dot*), the spatial-temporal information will be transferred to the matched testing STIPs (*blue triangle*) in the testing videos. By accumulating the votes from all the matching pairs, a subvolume is located in the *right* figure. The regions marked with *magenta* color refer to the low-dimension manifold learned with RPT, which can built on either training data or testing data (Best viewed in color)

with few query examples and (2) sufficient training data, e.g., in activity recognition with many training examples. The superior performances over the state-of-the-arts validate that our method can outperform in both conditions.

4.2 Activity Recognition by Detection

Similar to [6, 8], spatial-temporal interest points (STIP) [9] are first extracted from each video. For each STIP, we describe it with Histogram of Gradient and Histogram of Optical Flow. In total, the feature dimension is 162. We represent each training video with implicit spatial-temporal shape model based on extracted STIP points as shown in Fig. 4.1. Although we only apply sparse STIP feature in our experiments, our method is also applicable to dense local features. We refer to the training data as $\mathcal{R} : \{d_r = [\mathbf{f}_r, l_r]; r = 1, 2, \cdots, N_\mathcal{R}\}$, where $\mathbf{f}_r$ and l_r are the descriptor and 3D location of interest point d_r, respectively. $N_\mathcal{R}$ is the number of interest points.

Suppose we have a set of testing videos, denoted by $\mathcal{S} = \{\mathcal{V}_1, \mathcal{V}_2, \cdots, \mathcal{V}_{N_S}\}$, we want to recognize and locate the specific activity in the training video set $\mathcal{R}$, where $\mathcal{R}$ can be one or more training examples. Our goal is to find a video subvolume, V^*, to maximize the following similarity function:

$$\max_V S_{V \subset \mathcal{S}}(V, \mathcal{R}) = \max_{\mathbf{x}, t, \rho} S(V(\mathbf{x}, t, \rho), \mathcal{R}), \quad (4.1)$$

where $V(\mathbf{x}, t, \rho)$ refers to the subvolume with temporal center t and spatial center $\mathbf{x}$; ρ refers to the scale size and duration; $S(\cdot, \cdot)$ is the similarity measure. In total, we have 6 parameters (center position x, y, t, and width, height, duration) to locate the optimal subvolume V^*. For the problem of multi-class activity recognition (suppose we have K classes), Eq. 4.1 will be searched K times with training data $\mathcal{R}$ from different categories. Later, the class with the highest score of V^* will be picked as the predicting label for the testing video $\mathcal{V}$.

To measure the similarity between $V(\mathbf{x}, t, \rho)$ and the training data $\mathcal{R}$, similar to [10], we define $S(V(\mathbf{x}, t, \rho), \mathcal{R})$ in Eq. 4.1 as follows.

$$
\begin{aligned}
S(V(\mathbf{x}, t, \rho), \mathcal{R}) &= \sum_{d_r \in \mathcal{R}} p([\mathbf{x}, t, \rho], d_r) \\
&= \sum_{d_r \in \mathcal{R}} p([\mathbf{x}, t, \rho]|d_r)p(d_r),
\end{aligned}
\tag{4.2}
$$

where $d_r = [\mathbf{f}_r, l_r]$ with $\mathbf{f}_r$ representing the feature description and l_r representing the location of the rth STIP point in the training videos. $p([\mathbf{x}, t, \rho], \mathbf{f}_r, l_r)$ is the probability that there exists a target activity at position $[\mathbf{x}, t, \rho]$ and a matched STIP point d_r in the training data. Since it is reasonable to assume a uniform prior over d_r, we skip $p(d_r)$ and focus on the local feature voting $p([\mathbf{x}, t, \rho]|d_r)$:

$$
\begin{aligned}
p([\mathbf{x}, t, \rho]|d_r) &= \sum_{d_s \in \mathcal{S}} p([\mathbf{x}, t, \rho], d_s|d_r) \\
&= \sum_{d_s \in \mathcal{S}} p([\mathbf{x}, t, \rho]|d_s, d_r)p(d_s|d_r) \\
&= \sum_{d_s \in \mathcal{S}} p([\mathbf{x}, t, \rho]|l_s, l_r)p(\mathbf{f}_s|\mathbf{f}_r).
\end{aligned}
\tag{4.3}
$$

In Eq. 4.3, $p(\mathbf{f}_s|\mathbf{f}_r)$ determines the voting weight which relies on the similarity between $\mathbf{f}_s$ and $\mathbf{f}_r$. We will elaborate on how to compute $p(\mathbf{f}_s|\mathbf{f}_r)$ in Sect. 4.3. On the other hand, $p([\mathbf{x}, t, \rho]|l_s, l_r)$ determines the voting position. Suppose $d_r = [\mathbf{f}_r, l_r] \in \mathcal{R}$ matches $d_s = [\mathbf{f}_s, l_s] \in \mathcal{S}$, we cast the spatial-temporal information from the training data to the testing data with voting position $l_v = [\mathbf{x}_v, t_v]$:

$$
\begin{aligned}
\mathbf{x}_v &= \mathbf{x}_s - \eta_\mathbf{x}(\mathbf{x}_r - c_r^\mathbf{x}) \\
t_v &= t_s - \eta_t(t_r - c_r^t),
\end{aligned}
\tag{4.4}
$$

where $[\mathbf{x}_s, t_s] = l_s$, $[\mathbf{x}_r, t_r] = l_r$, $[c_r^\mathbf{x}, c_r^t]$ is the spatio-temporal center position of the training activity and $\eta = [\eta_\mathbf{x}, \eta_t]$ refers to the scale level and duration level (the scale size of the testing video, i.e., ρ, over the scale size of the matched training video). Once the voting position for testing sequence is available, we can compute $p([\mathbf{x}, t, \rho]|l_s, l_r)$ as:

$$
p([\mathbf{x}, t, \rho]|l_s, l_r) = \frac{1}{Z}e^{-\frac{\|[\mathbf{x}_v - \mathbf{x}, t_v - t]\|^2}{\sigma^2}},
\tag{4.5}
$$

where Z is a normalization constant and σ^2 is a bandwidth parameter.

4.3 Propagative Interest Point Matching

The matching of local features $p(\mathbf{f}_s|\mathbf{f}_r)$ plays an essential role in our Hough voting. According to Eq. 4.3, as each $d_r \in \mathcal{R}$ will be matched against all $d_s \in \mathcal{S}$, an efficient and accurate matching is essential. We propose to use the random projection trees [2] (RPT), which is constructed in an unsupervised way, to model the underlying low-dimension feature distribution, as the light magenta regions shown in Fig. 4.1. Compared with traditional Euclidean distance which ignores the hidden data distribution, RPT can give a more accurate evaluation of $p(\mathbf{f}_s|\mathbf{f}_r)$ with the help of underlying data distribution.

RPT has three unique benefits compared with other data structures, e.g., [11]. First of all, as proven in [2], random projection trees can adapt to the low-dimension manifold existing in a high dimension feature space. Thus, the matching found by random projection trees is superior to the nearest neighbor based on Euclidean distance. This advantage is further validated by our experimental results in Sect. 4.5. Second, similar to BoW model, we quantize the feature space by tree structures. Rather than enumerating all the possible interest point matches, we can efficiently find the matches by passing the query interest point from the root to the leaf nodes. This can save a lot of computational cost. Third, we can make more accurate estimation by increasing the number of trees. Later, we will prove that our random projection tree based Hough voting generates optimal solution when the number of trees approaches infinity. In the following section, we describe how to implement the random projection trees.

4.3.1 Random Projection Trees

Depending on the applications, our random projection trees can be built on (1) training data only, e.g., standard action classification and detection, (2) testing data only, e.g., activity search, and (3) both training and testing data. The trees are constructed in an unsupervised way and the labels from the training data will only be used in the voting step. Assume we have a set of STIPs, denoted by $\mathcal{D} = \{d_i; i = 1, 2, \cdots, N_{\mathcal{D}}\}$, where $d_i = [\mathbf{f}_i, l_i]$ as defined in Sect. 4.2 and $N_{\mathcal{D}}$ is the total number of interest points. The feature dimension is set to $n = 162$, so $\mathbf{f}_i \in R^n$.

We implement random projection trees [2] as shown in Algorithm 4. There are two parameters related to the construction of trees. N_T is the number of trees and δ_d is the maximum tree depth. Each tree can be considered as one partition of the feature space to index the interest points.

At the matching step, $p(\mathbf{f}_s|\mathbf{f}_r)$ in Eq. 4.3 will be computed as:

$$p(\mathbf{f}_s|\mathbf{f}_r) = \frac{1}{N_T} \sum_{i=1}^{N_T} I_i(\mathbf{f}_s, \mathbf{f}_r), \tag{4.6}$$

Algorithm 4 Trees = *ConstructRPT*($\mathcal{D}$)

1: **for** $i = 1 \rightarrow N_T$ **do**
2:    BuildTree($\mathcal{D}, 0$)
3: **end for**

4: **Proc** Tree = BuildTree($\mathcal{D}$, depth)
5: **if** depth $< \delta_d$ **then**
6:    Choose a random unit direction $v \in R^n$
7:    Pick any $x \in \mathcal{D}$; find the farthest point $y \in \mathcal{D}$ from x
8:    Choose γ uniformly at random in $[-1, 1] \cdot 6 \|x - y\| / \sqrt{n}$
9:    Rule(x) := $x \cdot v \leq (median(\{z \cdot v; z \in \mathcal{D}\}) + \gamma)$
10:    LTree $\leftarrow$ BuildTree($\{x \in \mathcal{D}; Rule(x) = true\}$, depth+1)
11:    RTree $\leftarrow$ BuildTree($\{x \in \mathcal{D}; Rule(x) = false\}$, depth+1)
12: **end if**

---

where N_T refers to the number of trees and

$$I_i(\mathbf{f}_s, \mathbf{f}_r) = \begin{cases} 1, & \mathbf{f}_s, \ \mathbf{f}_r \text{ belong to the same leaf in tree } T_i \\ 0, & \text{otherwise} \end{cases} \tag{4.7}$$

Thus, Eq. 4.2 becomes

$$\begin{aligned} S(V(\mathbf{x}, t, \rho), \mathcal{R}) &\propto \sum_{d_r \in \mathcal{R}} \sum_{i=1}^{N_T} \sum_{d_s \in \mathcal{S}} I_i(\mathbf{f}_s, \mathbf{f}_r) p([\mathbf{x}, t, \rho] | l_s, l_r) \\ &\propto \sum_{d_r \in \mathcal{R}} \sum_{i=1}^{N_T} \sum_{d_s \in \mathcal{S} \ \&\& \ I_i(\mathbf{f}_s, \mathbf{f}_r) = 1} p([\mathbf{x}, t, \rho] | l_s, l_r), \end{aligned} \tag{4.8}$$

where $d_s \in \mathcal{S} \ \&\& \ I_i(\mathbf{f}_s, \mathbf{f}_r) = 1$ refers to the interest points from $\mathcal{S}$ which fall in the same leaf as d_r in the *ith* tree. Based on Eq. 4.5, we can compute the voting score as:

$$S(V(\mathbf{x}, t, \rho), \mathcal{R}) \propto \sum_{d_r \in \mathcal{R}} \sum_{i=1}^{N_T} \sum_{d_s \in \mathcal{S} \ \&\& \ I_i(\mathbf{f}_s, \mathbf{f}_r) = 1} e^{-\frac{\|[x_v - \mathbf{x}, t_v - t]\|^2}{\sigma^2}}. \tag{4.9}$$

4.3.2 Theoretical Justification

The matching quality of Eq. 4.6 depends on the number of trees N_T. To justify the correctness of using random projection trees for interest point matching, we show that, when the number of trees is sufficient, our Hough voting algorithm can obtain the optimal detection results. For simplicity, we assume our hypothesis space is of size $W \times H \times T$, with W, H, T refer to the width, height and duration of the testing data, respectively. Each element refers to a possible center position for one activity and the scale ρ is fixed. We further assume there is only one target activity existing in the search space at the position $l^* = [\mathbf{x}^*, t^*]$. So in total there are $N_H = W \times H \times T - 1$ background positions. To further simplify the problem, we only vote for one position for each match rather than a smoothed region in Eq. 4.5. That is,

$$p(l^*|l_s, l_r) = \begin{cases} 1, & l^* = l_v \\ 0, & \text{otherwise} \end{cases} \tag{4.10}$$

We introduce a random variable $\mathbf{x}^{(i)}$ with Bernoulli distribution to indicate whether we have a vote for the position l^* or not in the ith match. We refer to the match accuracy as q and therefore $p(\mathbf{x}^{(i)} = 1) = q$. We introduce another random variable with Bernoulli distribution $\mathbf{y}^{(i)}$ to indicate whether we have a vote for the background position l_j (where $l_j \neq l^*$) or not in the ith match. Suppose each background position has an equal probability to be voted, then $p(\mathbf{y}^{(i)} = 1) = \frac{1-q}{N_H}$. We prove the following theorem in the supplementary material.

Theorem 4.1 *Asymptotic property of propagative Hough voting*: *When the number of trees $N_T \to \infty$, we have $S(V(l^*), \mathcal{R}) > S(V(l_j), \mathcal{R})$ with probability* $1 - \Phi(\frac{-(q - \frac{1-q}{N_H})\sqrt{N_M}}{\sigma_{xy}})$. *Specifically, if $q \geq \frac{1}{N_H+1}$, we have $S(V(l^*), \mathcal{R}) > S(V(l_j), \mathcal{R})$ when the number of trees $N_T \to \infty$.*

In Theorem 4.1, $\Phi(x) = \frac{1}{\sqrt{2\pi}} \int_{-\infty}^{x} e^{-\frac{x^2}{2}} dx$ and σ_{xy} refers to the variance. N_M refers to the number of matches according to Eq. 4.8: $N_M = N_T \times N_\mathcal{R} \times N_L$ if we build our RPT on the testing data, and $N_M = N_T \times N_S \times N_L$ if we build our RPT on the training data. N_L, referring to the average number of interest points in each leaf, can be estimated as $N_L \approx \frac{N_D}{2^{\delta_d}}$ where δ_d denotes the tree depth and N_D the size of the data for building RPT. Based on our empirical simulation experiments, q is much larger than $\frac{1}{N_H+1}$. Thus, the asymptotic property is true.

4.4 Scale Determination

To estimate ρ in activity localization, we propose an iterative refinement method, which iteratively applies the Hough voting and scale refinement. The reason we use the iterative algorithm is that we have six parameters to search for. This cannot be well handled in traditional Hough voting [10], especially when there is not sufficient amount of training data. We have two steps for the iterative refinement: (1) fix the scale, search for the activity center with Hough voting; (2) fix the activity center, and determine the scale ρ based on back-projection. We iterate the two steps until convergence.

The initial scale information ρ is set to the average scale of the training videos. Based on the Hough voting step discussed in Sect. 4.2, we can obtain the rough position of the activity center. Then back-projection, which has been used in [10, 13] for 2D object segmentation or localization, is used to determine the scale parameters.

After the Hough voting step, we obtain a back-projection score for each testing interest point d_s from the testing video based on Eq. 4.2:

$$
\begin{aligned}
s_{d_s} &= \sum_{d_r \in \mathcal{R}} p(l^* | l_s, l_r) p(\mathbf{f}_s | \mathbf{f}_r) \\
&= \frac{1}{Z} \sum_{d_r \in \mathcal{R}} e^{-\frac{\|l^* - l^s\|^2}{\sigma^2}} p(\mathbf{f}_s | \mathbf{f}_r),
\end{aligned}
\tag{4.11}
$$

where l^* is the activity center computed from last round; Z and σ^2 are, respectively, normalization constant and kernel bandwidth, which are the same as in Eq. 4.5. $p(\mathbf{f}_s | \mathbf{f}_r)$ is computed by Eq. 4.6. The back-projection score s_{d_s} represents how much this interest point d_s contributes to the voting center, i.e., l^*. For each subvolume detected in previous Hough voting step, we first enlarge the original subvolume in both spatial and temporal domains by 10%. We refer to the extended volume as $V^{l^*}_{W \times H \times T}$, meaning a volume centered at l^* with width W, height H and duration T. We need to find a subvolume $V^{l^*}_{w^* \times h^* \times t^*}$ to maximize the following function:

$$
\max_{w^*, h^*, t^*} \sum_{d_s \in V^{l^*}_{w^* \times h^* \times t^*}} s_{d_s} + \tau w^* h^* t^*,
\tag{4.12}
$$

where τ is a small negative value to constrain the size of the volume.

We assume each interest point which belongs to the detected activity would contribute in the Hough voting step, i.e., it should have a high back-projection score s_{d_s}. Thus, for those interest points with low back-projection scores, we consider them as the background. This motivates us to use the method in Eq. 4.12 to locate the optimal subvolume $V^{l^*}_{w^* \times h^* \times t^*}$.

Once we obtain the scale information of the subvolume, we replace ρ in Eq. 4.1 with (w^*, h^*, t^*) computed from Eq. 4.12 and start a new round of Hough voting. The process iterates until convergence or reaching to a pre-defined iteration number.

For activity classification, since the testing videos have already been segmented, the scale ρ can be determined by the width, height and duration of the testing video. The similarity between the training activity model and testing video defined in Eq. 4.1 is $S(V(\mathbf{x}^*, t^*, \rho), \mathcal{R})$ where $(\mathbf{x}^*, t^*)$ refers to the center position of the testing video.

4.5 Experiments

Two datasets are used to validate the performance of our algorithms. They are UT-Interaction [16] and TV Human Interaction dataset [12]. We perform two types of tests: (1) activity recognition with few training examples but we have a large testing data (building RPT on the testing data) and (2) activity recognition when the training data is sufficient (building RPT on the training data).

Table 4.1 Comparison of classification results on UT-Interaction (20 % training)

Method	[16]	[1]	NN + HV	RPT + HV
Accuracy	0.708	0.789	0.75	0.854

4.5.1 RPT on the Testing Data

In the following experiments, we first show that our algorithm is able to handle the cases when the training data is not sufficient. Experiments on UT-Interaction dataset and TV Human Interaction validate the performance of our algorithm. In these experiments, we build RPT using the testing data without labels. The reason why we do not train our RPT with both the training and testing data is that we need to handle the activity search problem, where we do not have the prior knowledge on query (training) data initially.

4.5.1.1 Activity Classification on UT-Interaction Dataset with 20 % Data for Training

We use the setting with 20 % data for training and the other 80 % for testing on UT-interaction dataset. This evaluation method has been used in [1, 16]. Since the training data is not sufficient, we build our random projection trees from the testing data. We list our results in Table 4.1. "NN + HV" refers to the method that nearest neighbor search is used to replace RPT for feature points matching. It shows that our algorithm has significant performance advantages compared with the state-of-the-arts.

4.5.1.2 Activity Classification on UT-Interaction Dataset with One Video Clip only for Training

Gaur et al. [5] provided the result of activity classification with training on only one video clip for each activity type and testing on the other video clips. To compare with [5], we performed the same experiments with just a single video clip as the training data for each activity type. We obtain an average accuracy of 73 %m, which is significantly better than the average accuracy of 65 % as reported in [5].

4.5.1.3 Activity Search with Localization on UT-Interaction Dataset

The activity search experiments are tested on the continuous UT-Interaction dataset. In the application scenario of activity search, there is usually just a few or even a single training sample available that indicates what kind of activity the user wants to find. Following the requirement of such an application scenario, we test our algorithm with only one query sample randomly chosen from the segmented videos. But if more training samples are available to our algorithm, the performance will be further boosted. With the help of our iterative activity search algorithm, we can efficiently locate all similar activities in a large un-segmented (continuous) video set. To compute the precision and recall, we consider a correct detection if: $\frac{\text{Volume}(V^* \cap G)}{\text{Volume}(V^* \cup G)} > \frac{1}{2}$ where G is the annotated ground truth subvolume, and V^* is the detected subvolume.

Figure 4.2 shows the results of different algorithms. The difference between our activity search and previous work is that we are only given one query video clip. Our system has no prior information about the number of activity categories in the database. In contrast to [5, 16], for every activity type, there is at least one video clip provided as training data. As previous works of activity search do not provide precision-recall curves, we only compare with the following algorithms: Branch and Bound [20, 23] (magenta curve) and nearest neighbors+Hough voting without scale determination (green curve). We use the same code provided by [20] to run the results. We list two categories of our results: (1) red curves: results after one step of Hough voting without scale refinement and (2) blue curves: results after one round of iteration (including both Hough voting and scale refinement). Compared with NN search, we can see the clear improvements by applying RPT to match feature points. Besides, back-projection refines the results from Hough voting. Since the dataset does not have very large spatial and temporal scale changes, we only present the results after one round of our iterative algorithm. The performance does not improve significantly when we further increase the number of iterations.

Figure 4.3 provides sample results of our activity search algorithm. One segmented video (sample frame for each category is shown in the first column) is used as the query and three detected results (marked with red rectangle) are included from the second to the forth column of Fig. 4.3.

4.5.1.4 Activity Search on TV Human Interaction Dataset

Since UT-Interaction is recorded in controlled environments, we use the TV Human Dataset [12] to show that our algorithm is also capable of handling activity recognition in uncontrolled environments. The dataset contains 300 video clips which are segmented from different TV shows. There are four activities: *hand shake, high five, hug* and *kiss*.

We have performed an experiment on the TV Human dataset for the performance evaluations on different number of training samples. We take the experiment with the following setting: 100 videos (25 videos for each category) as the database and randomly select a number of other videos as queries. RPT is built on the database (testing data). Figure 4.4 (Left) compares our results with those of NN + Hough voting. It shows the performance benefits of our RPT-based matching compared with nearest neighbor based matching.

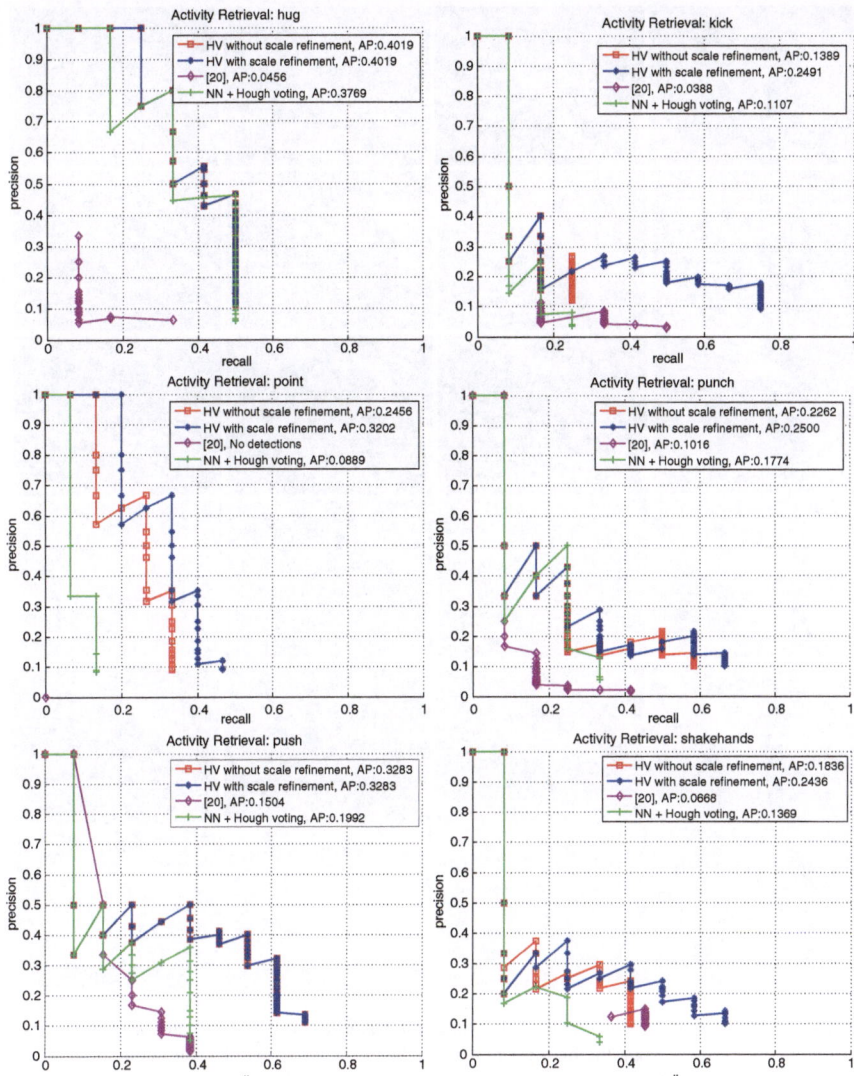

Fig. 4.2 Activity search results on UT-Interaction dataset

4.5.2 RPT on the Training Data

We have two experiments to further show that our algorithm can also have promising results for the traditional activity recognition problem, i.e., the training data is sufficient. One is tested on UT-Interaction dataset with Leave-one-out cross validation and another is on TV Human dataset.

Fig. 4.3 Activity search results on the UT-Interaction Dataset. We show one categories of results in each row. For each category, the first image is from the query video and the following three images are sample detection results. The *red* regions refer to our detected subvolumes

4.5.2.1 Activity Classification on UT-Interaction Dataset with Leave-One Validation

This setting was used in the activity classification contest [17]. It is a tenfold leave-one-out cross validation. Table 4.3 lists the published results on two different sets of videos. Since enough training data are provided, we build our unsupervised random projection trees from the training data without using the labels. The experimental

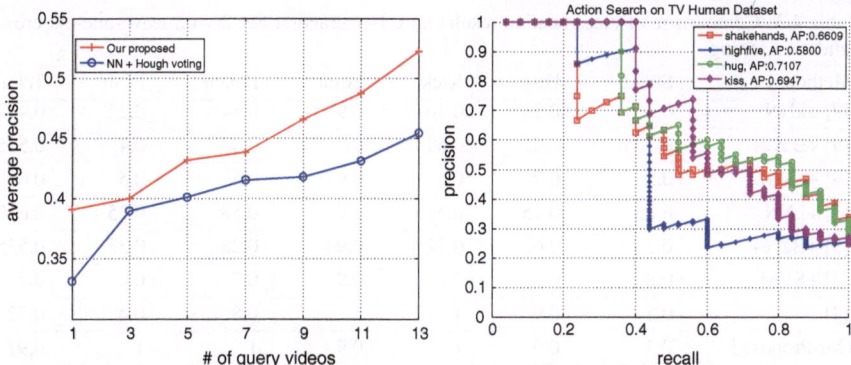

Fig. 4.4 *Left* Average precision versus different number of training videos provided on TV Human Dataset (testing on a database with 100 videos). *Right* PR curves for activity recognition on TV Human Dataset (25 videos for each activity used for training)

Table 4.2 Comparison of classification results on UT-Interaction **Set 1** with leave-one-out cross validation setting

Method	Shake	Hug	Kick	Point	Punch	Push	Total
[9] +kNN	0.18	0.49	0.57	0.88	0.73	0.57	0.57
[9] +Bayes	0.38	0.72	0.47	0.9	0.5	0.52	0.582
[9] +SVM	0.5	0.8	0.7	0.8	0.6	0.7	0.683
[3] +kNN	0.56	0.85	0.33	0.93	0.39	0.72	0.63
[3] +Bayes	0.49	0.86	0.72	0.96	0.44	0.53	0.667
[3] +SVM	0.8	0.9	0.9	1	0.7	0.8	0.85
[4]	0.7	1	1	1	0.7	0.9	0.88
[15] BoW	–	–	–	–	–	–	0.85
Our proposed	1	1	1	1	0.6	1	*0.933*

results show that our algorithm outperforms the state-of-the-art methods on the classification problem when the amount of training data is sufficient (Table 4.2).

As we can see from Table 4.3, results from cuboid features [3] are better than those from STIP features [9]. Even though we use STIP features, we still achieve better results than the state-of-the-art techniques that use cuboid features.

4.5.2.2 Action Recognition on TV Human Dataset

We test our algorithm using the standard setting as in [12]: training with 25 videos for each activity and testing on the remaining videos. In addition to [12], there are other works that published the results on this dataset. But they used additional information provided in the dataset, e.g., actor position, head orientation and interaction label of

Table 4.3 Comparison of classification results on UT-Interaction **Set 2** with leave-one-out cross validation setting

Method	Shake	Hug	Kick	Point	Punch	Push	Total
[9] +kNN	0.3	0.38	0.76	0.98	0.34	0.22	0.497
[9] +Bayes	0.36	0.67	0.62	0.9	0.32	0.4	0.545
[9] +SVM	0.5	0.7	0.8	0.9	0.5	0.5	0.65
[3] +kNN	0.65	0.75	0.57	0.9	0.58	0.25	0.617
[3] +Bayes	0.26	0.68	0.72	0.94	0.28	0.33	0.535
[3] +SVM	0.8	0.8	0.6	0.9	0.7	0.4	0.7
[4]	0.5	0.9	1	1	0.8	0.4	0.77
Our Proposed	0.7	0.9	1	0.9	1	1	0.917

Table 4.4 Comparison of activity classification on TV Human Dataset based on average precision

	100 videos	100 videos + 100 Neg
[12]	0.3933	0.3276
Our algorithm	0.6616	0.5595

each person. Thus, it is un-fair for us to compare with them since we only utilize the video data.

Following the evaluation method in [12], we also evaluate our algorithm based on average precision. Table 4.4 compares our results with those reported in [12]. "+Neg" means we add 100 negative videos that do not contain the target activities into the testing dataset. The precision-recall curves from our algorithm are shown in Fig. 4.4 (Right).

4.5.3 Computational Complexity

Here we only discuss the online computational cost as the RPT can be built offline. For the Hough voting step, it takes $O(N_M) + O(W'H'T')$, where N_M refers to the number of matches, which is defined in Sect. 4.3.2, and W', H', T' are the width, height and duration of the testing videos, respectively. For the back-projection step, the computational complexity is $O(N_M) + O(WHT)$, where W, H, T are the width, height and duration of the extended subvolume defined in Sect. 4.4 and $T << T'$. It takes approximately 10 s to perform the activity classification for each 4-s long testing video and 15 s for activity search on a 1 min testing video on the UT-Interaction dataset. The feature extraction takes a few more seconds depending on the length of the video. The system is implemented in C++ and runs on a regular desktop PC.

4.6 Summary of this Chapter

Local feature voting plays an essential role in Hough voting-based detection. To enable discriminative Hough-voting with limited training examples, we propose propagative Hough voting for human activity analysis. Instead of matching the local features with the training model directly, by using random projection trees, our technique leverages the low-dimension manifold structure in the high-dimensional feature space. This provides us significantly better matching accuracy and better activity detection results without increasing the computational cost too much. As the number of trees grows, our propagative Hough voting algorithm can converge to the optimal detection. The superior performances on two benchmarked datasets validate that our method can outperform not only with sufficient training data, e.g., in activity recognition, but also with limited training data, e.g., in activity search with one query example.

References

1. W. Brendel, S. Todorovic, Learning spatiotemporal graphs of human activities, *ICCV* (2011)
2. S. Dasgupta, Y. Freund, Random projection trees and low dimensional manifolds, *ACM symposium on Theory of computing (STOC)*, pp. 537–546 (2008)
3. P. Dollar, V. Rabaud, G. Cottrell, S. Belongie, Behavior recognition via sparse spatio-temporal features. Workshop on Visual Surveillance and Performance Evaluation of Tracking and Surveillance (2005)
4. J. Gall, A. Yao, N. Razavi, L. Van Gool, V. Lempitsky, Hough forests for object detection, tracking, and action recognition, in *IEEE Transaction on Pattern Analysis and Machine Intelligence*, pp. 2188–2202 (2011)
5. U. Gaur, Y. Zhu, B. Song, A. Roy-Chowdhury, A string of feature graphs model for recognition of complex activities in natural videos, in *ICCV* (2011)
6. N.A. Goussies, Z. Liu, J. Yuan, Efficient search for top-K video subvolumes for multi-instance action detection, in *IEEE International Conference on Multimedia and Expo* (2010)
7. Y. Ke, R. Sukthankar, M. Hebert, Event detection in crowded videos, in *Proceedings of the IEEE International Conference on Computer Vision*, pp. 1–8 (2007)
8. I. Laptev, P. Prez, Retrieving actions in movies, in *Proceedings of the ICCV* (2007)
9. I. Laptev, On space-time interest points. Int. J. Comput. Vis. **64**(2–3), 107–123 (2005)
10. B. Leibe, A. Leonardis, B. Schiele, Robust object detection with interleaved categorization and segmentation. IJCV **77**(1–3), 259–289 (2007)
11. F. Moosmann, E. Nowak, F. Jurie, Randomized clustering forests for image classification. PAMI **30**, 1632–1646 (2008)
12. A. Patron-perez, M. Marszalek, A. Zisserman, I. Reid, High five: recognising human interactions in TV shows, *BMVC* (2010)
13. N. Razavi, J. Gall, L. Van Gool, Backprojection revisited: scalable multi-view object detection and similarity metrics for detections, in *ECCV* (2010)
14. K.K. Reddy, J. Liu, M. Shah, Incremental action recognition using feature-tree, in *Proceedings of the IEEE International Conference on Computer Vision*, pp. 1010–1017 (2009)
15. M.S. Ryoo, Human activity prediction: early recognition of ongoing activities from streaming videos, in *ICCV* (2011)
16. M.S. Ryoo, J.K. Aggarwal, Spatio-temporal relationship match: video structure comparison for recognition of complex human activities, in *ICCV* (2009)

17. M.S. Ryoo, C. Chen, J. Aggarwal, An overview of contest on semantic description of human activities (SDHA), in *SDHA* (2010)
18. C. Schuldt, I. Laptev, B. Caputo, Recognizing human actions: a local svm approach, in *Proceedings of the IEEE Conference on Pattern Recognition*, pp. 57–62 (2004)
19. G. Yu, J. Yuan, Z. Liu, Real-time human action search using random forest based hough voting, *ACM Multimedia* (2011)
20. G. Yu, J. Yuan, Z. Liu, Unsupervised random forest indexing for fast action search, in *CVPR* (2011)
21. G. Yu, A. Norberto, J. Yuan, Z. Liu, Fast action detection via discriminative random forest voting and top-K subvolume search. IEEE Trans. Multim. **13**(3), 507–517 (2011)
22. G. Yu, J. Yuan, Z. Liu, Action search by example using randomized visual vocabularies. IEEE Trans. Image Process. **22**(1), 377–390 (2013)
23. J. Yuan, Z. Liu, Y. Wu, Discriminative video pattern search for efficient action detection. IEEE Trans. Pattern Anal. Mach. Intell. **33**, 1728–1743 (2011)

Chapter 5
Human Action Prediction with Multiclass Balanced Random Forest

Abstract Early recognition and prediction of human activities are of great importance in video surveillance. In this chapter, we target this problem by developing a spatial-temporal implicit shape model (STISM), which characterizes the space-time structure of the sparse local features extracted from a video. The recognition of human activities is accomplished by pattern matching through STISM. To enable efficient and robust matching, we propose a new random forest structure, called multiclass balanced random forest, which makes a good trade-off between the balance of the trees and the discriminative abilities. The prediction is done simultaneously for multiple classes, which saves both the memory and computational cost. The experiments show that our algorithm significantly outperforms the state-of-the-art for the human activity prediction problem.

Keywords Activity prediction · Multiclass balanced random forest · Partial observation · Spatial-temporal implicit shape model · Multiclass recognition

5.1 Introduction

In Chaps. 2–4, we presented three algorithms for human action analysis based on complete video observations. In this chapter, we will address an efficient and effective algorithm for human action prediction based on uncomplete video observations.

The problem of human activity prediction has been proposed in [4]: *inference of the ongoing activity given temporally incomplete observations*. For instance, in a supermarket, it is better to send off an alarm while someone is stealing rather than after the stealing, because it can possibly prevent this criminal activity and also provide more time for the security guard to react. As another example shown in Fig. 5.1, when there are people fighting on a street, it is extremely useful to recognize and stop the fighting activity early before the situation becomes worse. Integral bag-of-words (BoW) and dynamic bag-of-words are proposed in [4] to enable activity prediction with only partial observations. Despite certain successes of [4], it still has several limitations. First, since the BoW model ignores the spatial-temporal relationships

© The Author(s) 2015

G. Yu et al., *Human Action Analysis with Randomized Trees*,
SpringerBriefs in Signal Processing, DOI 10.1007/978-981-287-167-1_5

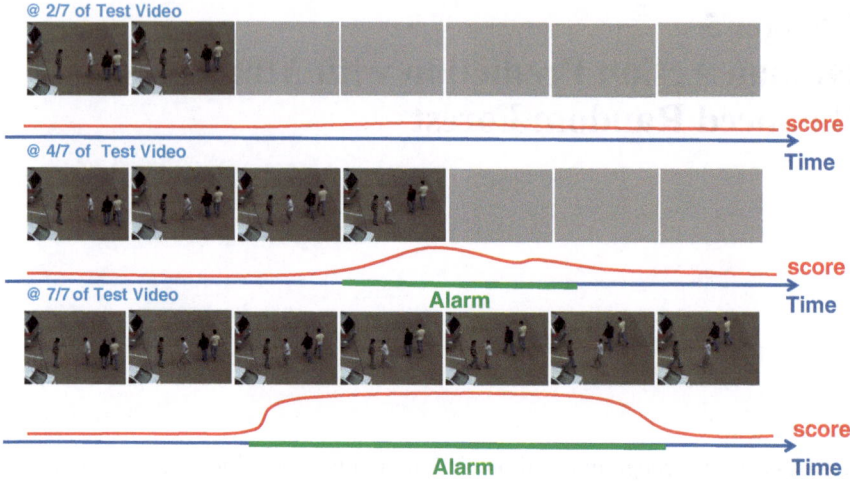

Fig. 5.1 An illustration of the human activity prediction problem. We want to predict the "push" activity and seven sample frames are selected among the testing video. Three experiments on different observation ratios (at 2/7, 4/7, 7/7) are shown with the *red curve* describing the score at each frame. The *green solid line* on the time coordinate refers to the predicted activity in the testing video

among interest points, it is not discriminative enough to describe human activities. Also, although integral BoW and dynamic BoW in [4] consider the temporal information by matching between subintervals, there lacks a principled way to determine the optimal interval length. Finally, as we usually have a large number of categories of activities to detect, it demands an algorithm whose computational complexity is sublinear or constant to the number of categories.

To address these limitations, we utilize Spatial-Temporal Implicit Shape Model (STISM), which is proposed in Chap. 4 to model the relationships between the local features. To enable efficient and robust matching, a new type of random forest is proposed, which makes a good trade-off between the tree balance and discriminative ability. Meanwhile, the trees will be trained for multiclass purpose, which makes our algorithm scalable to the number of classes. Given a normal desktop PC, our human activity prediction algorithm can be run in real-time. In addition to the speed benefit, STISM makes it possible to progressively predict the human activities thanks to the additive nature of the model. Even when we only have partial observation, the prediction can be accurate as well. Our action prediction experiments on UT-Interaction dataset [5] further validate the performance of our algorithm.

5.2 Problem Formulation

We represent the videos with spatial-temporal interest point (STIP) [1] due to its sparsity and good performance for action recognition. Other types of local features are also applicable to our algorithm. Given a video, we refer our STISM as $\mathcal{V} =$

$\{(f_i, s_i, c)\}$, where f_i refers to the feature description, $s_i = l_i - l_\mathcal{V}$ refers to the spatio-temporal location shift from the ith STIP position (l_i) to the center position of video $l_\mathcal{V}$, and c refers to the category of the video. STISM is a 3D extension of implicit shape model in [2]. The detailed information about STISM can be referred to [7]. Figure 5.2 illustrates the idea of using STISM for activity matching. The yellow dots refer to the detected interest points and the white dash lines refer to the shift from the interest point to the video center, i.e., $l_i - l_\mathcal{V}$. The benefits of our implicit shape model are twofold. On one hand, it is flexible to utilize the spatial-temporal configuration of the interest points for recognition. More specifically, we do not need to explicitly define and learn a model. On the other hand, the computational cost is low which enables real-time activity prediction. Our goal is, given a training set $\mathcal{D} = \{(f_j, s_j, c_j)\}$ (several different f_j will share the same video center location and c_j if they are from the same training video), to determine the category of c for testing video $\mathcal{V}$. Following [2], our similarity score of an incomplete testing video $\mathcal{V}^\delta$ belonging to a specific class $C \in \{1, 2, \ldots, K\}$ is defined as:

$$
\begin{aligned}
S(C, \mathcal{V}^\delta, l_\mathcal{V}) &= \sum_{(f_i, l_i) \in \mathcal{V}^\delta} p(c_i = C, l_\mathcal{V}, f_i, l_i) \\
&\propto \sum_{(f_i, l_i) \in \mathcal{V}^\delta} p(c_i = C, l_\mathcal{V} | f_i, l_i),
\end{aligned}
\tag{5.1}
$$

where $\mathcal{V}^\delta$ refers to the percentage ($\delta \in [0, 100\,\%]$) of video $\mathcal{V}$ observed. The prior $p(f_i, l_i)$ in Eq. 5.1 is assumed to follow a uniform distribution. The probability of $p(c_i = C, l_\mathcal{V} | f_i, l_i)$ can be computed as:

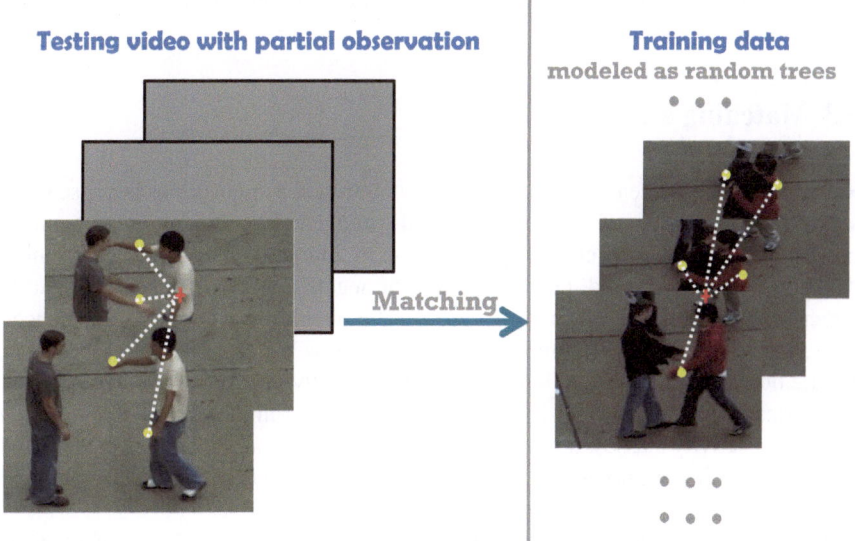

Fig. 5.2 An illustration of our spatio-temporal activity matching

$$p(c_i = C, l_V | f_i, l_i) = \sum_{(f_j, s_j, c_j = C) \in \mathcal{D}} p(c_i = C, l_V | f_j, s_j, c_j = C, f_i, l_i)$$

$$\times p(f_j, s_j, c_j = C | f_i, l_i) \qquad (5.2)$$

$$= \sum_{(f_j, s_j, c_j = C) \in \mathcal{D}} p(c_i = C, l_V | s_j, c_j = C, l_i)$$

$$\times p(f_j, c_j = C | f_i).$$

Similar to [2], we made two assumptions for Eq. 5.2. The first assumption is:

$$p(c_i = C, l_V | f_j, s_j, c_j = C, f_i, l_i) = p(c_i = C, l_V | s_j, c_j = C, l_i),$$

referring to the similarity based on the spatial-temporal shifts (white dash line in Fig. 5.2). We can further compute it as:

$$p(c_i = C, l_V | s_j, l_i, c_j = C) = \frac{1}{Z} \exp^{\frac{-((l_i - l_V) - s_j))^2}{\sigma^2}}, \qquad (5.3)$$

where Z is a normalization constant and σ^2 is a bandwidth parameter. The second assumption is that $p(f_j, s_j, c_j = C | f_i, l_i) = p(f_j, c_j = C | f_i)$, which serves as a weight based on feature description for each matched interest point pair $(f_j, s_j, c_j) \in \mathcal{D}$ and $(f_i, l_i) \in \mathcal{V}$.

To reduce the computational cost caused by the enumeration of all the interest point pairs in Eq. 5.2, we propose a new random forest structure, called *Multiclass Balanced Random Forest* (MBRF), in the next section. With the help of MBRF, we only need to focus on the interest point pairs which fall into the same leaf.

5.3 Matching and Predicting

Random forest has been widely used in many multimedia applications because it has superior performance and fast computational speed. However, for action recognition problems, there are two challenges we need to address. First, since our training data is usually unbalanced, i.e., the number of negative videos will be significantly larger than the number of positive videos. This would easily lead to unbalanced trees, resulting in low discriminative ability and low matching accuracy. Besides, for the human activity prediction problem, one usually needs to recognize multiple categories of actions. It is desirable to develop an algorithm that is scalable to the number of activity classes. Instead of building one-versus-all random forest for each category as in [8], we use a multiclass random forest that saves a lot of computation and storage.

Given the training data, $\mathcal{D} = \{(f_j, s_j, c_j), j = 1, 2, \ldots, N_{\mathcal{D}}\}$ where f_j is described with Histogram of Gradient (HoG) and Histogram of Flow (HoF), we construct N_T trees as follows. For each node, we choose one of the two splitting measures with equal probability:

- Distribution based measure: to ensure the tree balance and model the underlying data distribution.
- Entropy based measure: to ensure the discriminative ability of the trees.

N_h hypotheses will be generated for each node based on the selected splitting measure. For each hypothesis with the distribution based measure, we randomly select two dimension indexes τ_1 and τ_2 (either from HoG or HoF part). The variance of the training data on the two dimensions can be computed:

$$\text{Var}_{(\tau_1,\tau_2)} = \sum_{j=1}^{N_D}((f_j(\tau_1) - f_j(\tau_2)) - \mu_{(\tau_1,\tau_2)})^2, \tag{5.4}$$

where $\mu_{(\tau_1,\tau_2)} = \frac{1}{N_D}\sum_{j=1}^{N_D}(f_j(\tau_1) - f_j(\tau_2))$.

Based on the N_h hypotheses, we select the one with the largest variance $\text{Var}_{(\tau_1,\tau_2)}$ and the corresponding $\mu_{(\tau_1,\tau_2)}$ is used as the splitting threshold. This distribution based splitting measure has two benefits. On one hand, it can make the trees balanced since the two child nodes after splitting are usually of the similar size. On the other hand, this can be considered as the data distribution modeling. Consider the extreme case when all the nodes are split with distribution based measure, the tree constructing process can be considered as a clustering step, with each leaf as a word in a vocabulary (BoW). The difference between our random forest with the BoW is that we have multiple trees and the variance of the estimation error can be reduced.

Another splitting measure is entropy based measure. Similarly, we generate a set of hypotheses. For each hypothesis, two random numbers for feature dimension indexes are first generated, τ_1 and τ_2. A small jitter value, ξ, is randomly generated as well. We set splitting threshold γ as:

$$\gamma = \frac{1}{N_D}\sum_{j=1}^{N_D}(f_j(\tau_1) - f_j(\tau_2)) + \xi. \tag{5.5}$$

The node will be split into two child nodes based on the threshold γ. For the current node, we can compute the entropy as

$$\begin{aligned}E(\tau_1, \tau_2, \gamma) = &\frac{1}{|F_l|}\sum_{C=1}^{K} -p_C(F_l)\log(p_C(F_l)) \\ &+ \frac{1}{|F_r|}\sum_{C=1}^{K} -p_C(F_r)\log(p_C(F_r)),\end{aligned} \tag{5.6}$$

where

$$F_l = \{j : f_j(\tau_1) - f_j(\tau_2) < \gamma\}$$
$$F_r = \{j : f_j(\tau_1) - f_j(\tau_2) \geq \gamma\},$$

(5.7)

and

$$p_C(F_l) = \frac{1}{|F_l|} \sum_{j \in F_l} I(c_j = C),$$

(5.8)

is the probability of samples belonging to the category c in the F_l set. $I(x)$ is an identity function. We can define $p_C(F_r)$ in a similar way. By choosing the hypothesis with the smallest entropy, we can increase the discriminative ability of our trees.

The above process is repeated until the predefined maximum tree depth is reached or the number of feature points in a node is smaller than a pre-defined number. We name this tree structure as Multiclass Balanced Random Forest. Now let us revisit Eq. 5.2. The weight for the interest point pair can be computed as:

$$p(f_j, c_j = C|f_i) = \frac{1}{|N_T|} \sum_{t=1}^{N_T} \frac{1}{|L_t|} \sum_{j \in L_t} I(c_j = C),$$

(5.9)

where L_t refers to the leaf node in tth tree which both training STIP j and testing STIP i fall in. Rather than enumerating all the interest point pairs between training data and testing video, the computational cost can be significantly reduced by traversing our MBRF and a small subset of interest points from the training data $\mathcal{D}$ will be found with positive weight $p(f_j, c_j = C|f_i)$ for each testing interest point i.

The final score for the segmented testing video is the accumulation of all the votes based on Eq. 5.1. The category of the video given observation $\mathcal{V}^\delta$ is then determined by:

$$C^* = \arg\max_C S(C, \mathcal{V}^\delta, l_\mathcal{V}).$$

(5.10)

With more observations δ provided, more matched pairs will be found and the score will be increased. Thus, the results will be further refined.

In our method, the scale (both spatial and temporal) variations are ignored based on two reasons. First, since our STIP feature already encodes the scale information, the matched STIP pair i and j share the similar activity scale. Second, we add a smooth kernel in Eq. 5.3 for each vote so that the small scale variations can be well handled.

5.4 Experiments

We choose UT-Interaction dataset [5] to evaluate our algorithm for the following two reasons. First, the dataset is recorded under the realistic surveillance environment. Second, the activities that a surveillance system is interested in predicting such as

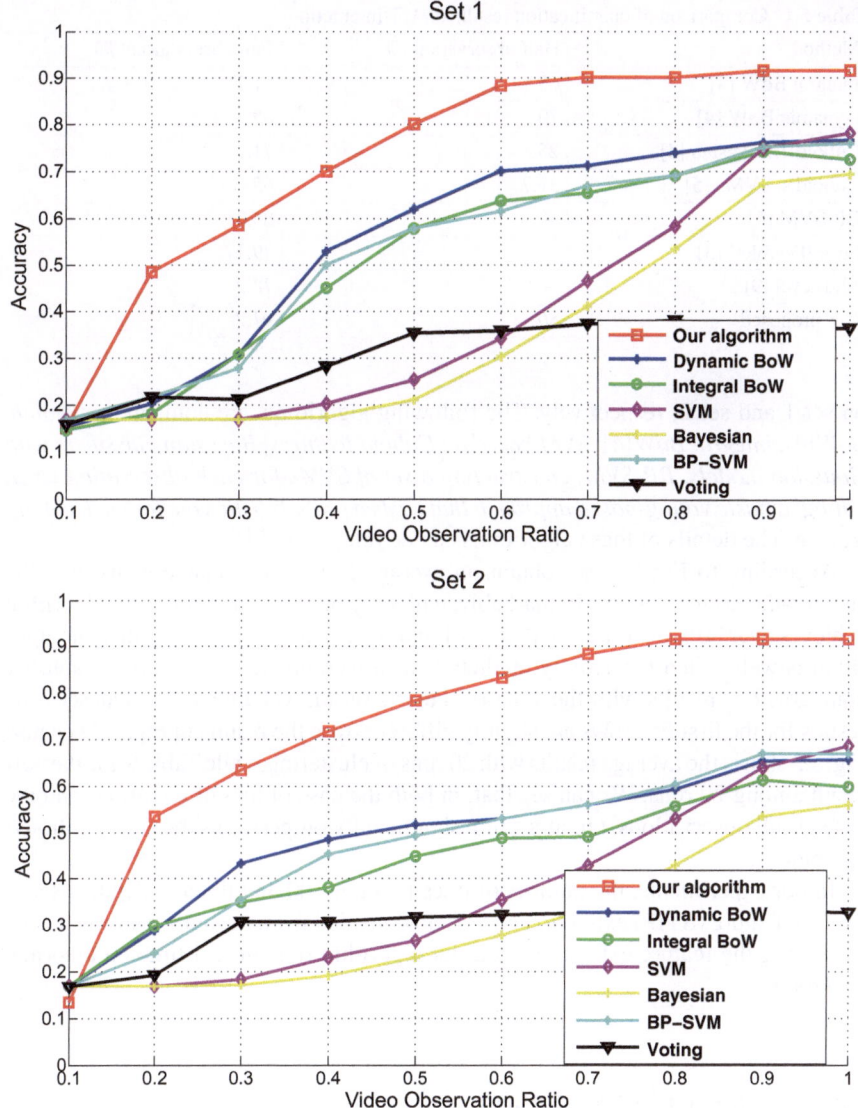

Fig. 5.3 Human activity prediction on UT-Interaction dataset. (Set 1: *Up*; Set 2: *Down*)

shoplifting are usually nonperiodic and instantaneous, similar to the six activities in UT-interaction dataset. UT-interaction dataset contains two scenes with 60 videos each. The six types of activities are: *handshaking, pushing, punching, pointing, kicking*, and *hugging*.

The same setting as [4] (Leave-one sequence-out cross validation) is used to evaluate our algorithm. Figure 5.3 shows the results compared with the other algorithms

Table 5.1 Comparison of classification results on UT-Interaction

Method	Half observation (%)	Full observation (%)
Integral BoW [4]	65	81.7
Dynamic BoW [4]	70	85
Cuboid + Bayesian [4]	25	71.7
Cuboid + SVMs [5]	31.7	85
BP-SVM [6]	–	83.3
Pose 'Doublet' [3]	–	79.17
Mid-level [9]	–	78.2
Our proposed	*80*	*91.7*

on set 1 and set 2, respectively. The following algorithms are compared: *dynamic BoW*[4], *integral BoW*[4], *SVM based on Cuboid features, Bayesian classifiers with Gaussian models, BP-SVM: constructing a set of SVMs for each observation level, Voting: a basic voting-based approach that casts a probabilistic vote for each cuboid feature.* The details of these algorithms can be referred to [4].

According to Fig. 5.3, we obtain on average 20 % performance gains over the state-of-the-art techniques. Remarkably, with only 60 % observations, our algorithm achieves over 80 % accuracy on both set 1 and set 2. This demonstrates that our algorithm is well suited for activity prediction with incomplete observations. Table 5.1 compares our results with the state-of-the-arts on the UT-Interaction dataset. The results for the first five rows are slightly different from the results in Fig. 5.3 because Fig. 5.3 shows the average results with 20 runs of clustering while Table 5.1 is the best result among 20 runs. We can see that, in both the case of half observations and the case of full observations, our algorithm significantly outperforms the state-of-the-art techniques.

In our experiments, the number of trees is set to 100 and the tree depth is set to 15. The feature (STIP) extraction code is downloaded from the author's website [1]. Excluding the feature extraction cost, our algorithm runs in real-time on a normal desktop PC.

5.5 Summary of this Chapter

In this chapter, we presented a simple yet surprisingly effective solution for human activity prediction problem. Spatial-temporal implicit shape model is utilized to capture the spatio-temporal structure of local features. Matching between the testing and training video is effectively and efficiently solved with our proposed multiclass balanced random forest, which makes a good trade-off between the discriminative ability and tree balance. Besides, our MBRF models all classes simultaneously and therefore is scalable to multiclass prediction. Experimental results show that our algorithm significantly outperforms the state-of-the-arts. In future work, we plan to handle the activity prediction problem on unsegmented videos.

References

1. I. Laptev, On space-time interest points. Int. J. Comput. Vis. **64**(2–3), 107–123 (2005)
2. B. Leibe, A. Leonardis, B. Schiele, Robust object detection with interleaved categorization and segmentation. IJCV **77**(1–3), 259–289 (2007)
3. S. Mukherjee, S.K. Biswas, D.P. Mukherjee, Recognizing interaction between human performers using 'Key Pose Doublet', *ACM Multimedia* (2011)
4. M.S. Ryoo, Human activity prediction: early recognition of ongoing activities from streaming videos, in *ICCV* (2011)
5. M.S. Ryoo, C. Chen, J. Aggarwal, An overview of contest on semantic description of human activities (SDHA), *SDHA* (2010)
6. T.-H. Yu, T.-K. Kim, R. Cipolla, Real-time action recognition by spatiotemporal semantic and structural forests, *BMVC* (2010)
7. G. Yu, J. Yuan, Z. Liu, Propagative hough voting for human activity recognition, *Proceedings of the European Conference on Computer Vision (ECCV)* (2012)
8. G. Yu, A. Norberto, J. Yuan, Z. Liu, Fast action detection via discriminative random forest voting and top-K subvolume search. IEEE Trans. Multim. **13**(3), 507–517 (2011)
9. F. Yuan, V. Prinet, J. Yuan, Middle-level representation for human activities recognition: the role of spatio-temporal relationships, in *ECCV Workshop on Human Motion* (2010)

Chapter 6
Conclusion

Abstract Generally, this book provides a systematic study on the human action analysis problems based on tree-based approaches. This chapter concludes the four approaches presented in this book.

Keywords Action recognition · Action detection · Action search · Action prediction · Tree based approach

To analyze human actions in videos, we proposed four effective and efficient algorithms for different human action analysis tasks. In Chap. 2, we presented a random forest-based template matching method to detect actions, which significantly improve the speed issue for human action recognition and detection with the help of spatial downsampling and Top-K search. In Chap. 3, the speed is further improved with the help of coarse-to-fine branch and bound search, which makes it possible for real-time action search. Besides, unsupervised random indexing tree is used to implement the multiple visual vocabularies, which can improve the matching accuracy of the local interest points. In Chap. 4, we proposed Hough voting-based approach for human action recognition and action search search. Unsupervised random projection trees is utilized to leverage the underlying data distribution. With the help of propagative Hough voting, the local interest point matching is improved by the spatio-temporal configuration of local interest points. Following the framework in Chap. 4, we proposed multiclass balanced random forest to handle the action prediction problem in Chap. 5.

© The Author(s) 2015
G. Yu et al., *Human Action Analysis with Randomized Trees*,
SpringerBriefs in Signal Processing, DOI 10.1007/978-981-287-167-1_6